SUCCESSFUL ALTERNATE ENERGY METHODS

⌂ SUCCESSFUL ALTERNATE ENERGY METHODS

James D. Ritchie

Structures Publishing Company
Farmington, Michigan 48024

Manufactured in the United States of America

Edited by Peggy Frohn

Produced by Carey Jean Ferchland

Text Design by Linda A. Zitzewitz

Cover photos courtesy of (*Top left*) Freelance Photographers Guild, J. Zimmerman; (*Top center*) Bror Hanson, Solargy Corporation; (*Top right*) Freelance Photographers Guild, J. Zimmerman; (*Bottom*) Freelance Photographers Guild

Current Printing (last digit)
10   9   8   7   6   5   4   3   2   1

Structures Publishing Co.
24277 Indoplex Circle
Box 1002, Farmington, Mich. 48024

Library of Congress Cataloging in Publication Data

Ritchie, James D
  Successful alternate energy methods.

  Bibliography: p.
  Includes index.
  1. Dwellings—Energy conservation. 2. Renewable energy sources. I. Title.
TJ163.5.D86R57   696   80-13103
ISBN 0-89999-000-2
ISBN 0-89999-001-0 (pbk.)

CONTENTS

This experimental wind turbine, built by the U.S. Department of Energy and National Aeronautics and Space Administration engineers, stands at Block Island, Rhode Island. The 125-feet-diameter rotor develops 200 kilowatts of power in an 18 m.p.h. wind. (Courtesy of NASA)

INTRODUCTION

"Today the need to develop and expand renewable energy sources that can provide heating, cooling and power for homes, farms and factories is greater than at any other time in our nation's history."

President Jimmy Carter
March 27, 1978

Is there *really* an energy crisis in America?

President Carter has said that our energy situation is the "moral equivalent of war," but ran into trouble convincing the U.S. Congress to declare war on our energy problems.

Whether the fuel shortage has reached crisis proportions is being argued not only in Washington, D.C., but across the country. However, no one who has observed what is happening in America can deny that we have serious energy *problems.*

When gasoline prices increase more than 50 percent in less than a year—that's a *problem.* When home-heating fuel goes up an average of four or five percent per month, to the point that many low-income families have to choose between heating and eating—that's a *problem.* When the U.S. imports about $165 million worth of foreign oil each day, with no certain national policy in sight to turn down the spigot—we've all got problems.

It's a real problem for homeowners, apartment dwellers and businessmen who must buy energy at ever inflated prices.

"Our trouble isn't just an energy shortage, it's more critically a shortage of energy in the forms we need to do the job," says Dr. William Hughes, electrical engineer, Oklahoma State University. "Half of the high-grade energy we use—oil, gas and electricity—goes to heat living spaces and water.

"We need to use alternate resources to heat homes, offices, factories and hot water, to release our high-grade energy for those jobs that only high-grade energy can do," says Dr. Hughes. "Low-grade" energy can be supplied by solar power, wood burning or geothermal wells; sources that cannot handily be used to operate motor vehicles or power electrical equipment. If U.S. homes, factories and businesses were heated by "low-grade" sources of energy, half of the gas, oil and electricity we now use could be reserved to power vehicles and factories, to operate lights and electrical equipment. In other words, we'd have twice as much "high-grade" energy to perform those functions that only high-grade energy can perform.

"We need more money put into research and development of alternate energy sources, but we need good ideas even more," says Dr. Hughes. "For example, we need new ideas on how to capture and use waste heat from electrical power generation and manufacturing processes; we need more practical application of some of the newer technology being developed.

"We have to conserve our high-grade sources of energy to keep the economic plant running," he adds. "But any culture that is dependent on the resources of other people is bound to change, and we are importing a great deal of oil from abroad. Our life styles will have to become more basic, simpler. We'll walk more and drive less. We'll dress differently. We'll build our homes and public buildings differently—probably better. The change may be painful for awhile, but it surely doesn't have to be fatal."

The change is already underway, and those Americans in the vanguard are finding it relatively painless. President Carter's "moral equivalent of war" may be stalemated in Washington, but private citizens, public utilities and local governments from coast to coast are winning victories in the first skirmishes. From New York to San

Diego, it's American ingenuity at its best.

Here are a few examples of what is being done:

- The University of Wisconsin at Milwaukee saves $80,000 per year in electric bills by pumping cool water from nearby Lake Michigan through campus air-conditioning systems. The school not only saves money, but maintains classroom temperatures at 70 degrees —eight degrees cooler than the federally-mandated summer setting.
- Residents of Burlington, Vermont, recently voted a bond issue to finance construction of a 50-megawatt generator that is to be powered by burning wood.
- Minnesota dairy farmer, Lance Crombie, built a solar-heated evaporative still to produce fuel-grade grain alcohol to run motor vehicles and farm equipment.
- Solar energy is a community affair for Davis, California, where solar homes have been constructed since 1973 in entire subdivision developments—with the result that natural gas consumption has dropped by more than a third.
- City-owned power plants at Ames, Iowa, burn trash and other refuse to generate electricity for nearly 70,000 customers.
- In Klamath Falls, Oregon, and Boise, Idaho, the hot breath of Mother Earth from geothermal wells heats homes and public buildings, at a savings of some 500,000 barrels of oil per year.
- The Missouri Power and Light Company, a public utility that serves 35,000 homes with natural gas, had not signed up any new customers since the 1973 Arab oil boycott—until 1979, when the utility started hooking up new gas customers. The company hadn't found any new gas supplies—those original 35,000 customers are using gas more efficiently now, which allows that limited source of energy to be used by more people.
- Odell Morgan, of Norman, Oklahoma, attached a solar greenhouse to the south wall of his home, and saved a third on heating bills. "We not only saved money on heating fuel, the growing plants in the greenhouse added humidity to the air throughout the house, and my family has been bothered less with colds and sniffles in winter," says Morgan.

The above are only a few examples of what is happening all over the country, as Americans cook up ingenious ways to produce and save energy—and money. The Internal Revenue Service reports nearly six million American taxpayers claimed more than $4 billion worth of energy saving devices on their tax returns in 1977-78. Obviously, American people aren't waiting for Washington to agree on how serious the energy problem is or what should be done about it.

Wherever you live, you can start easing your way into the natural energy picture right now. If you look through the above list of things individuals, governments and organizations are doing, you'll notice that most of what is being done in this whole area of alternate energy is neither very new nor very complicated. The sun has been there forever. People have burned wood for heat for a long time. We've known about the potential heat in geysers and underground "hot" wells for years.

Perhaps it would make more sense to call the sun, wood, wind, water and earth *natural* sources of energy, and describe petroleum as the *alternate* source. After all, we went to widespread use of gas and oil because that alternative was very convenient to use—and for a time, very cheap. Now, with each jump in petroleum prices, that alternative is less attractive. And each price increase in oil, gas or petroleum generated electricity brings more people back to natural sources of energy.

Before you read any further, dig out the energy bills (or cancelled checks) you paid during the last 12 months. Round up all of them—electricity, gas, fuel oil, coal or whatever—everything you paid someone else for heat and power. Now, total up the cost of energy for the entire year. Chances are, it's a sizeable sum. That's your potential savings—right now. How much you will be able to save on energy bills in the future, as inflation and ever scarcer fuel supplies push up the cost, is anybody's guess.

Wouldn't it be wonderful to live in a solar-heated home, with no heating bills to pay? Or have a windmill on the hill to generate all the "free" electricity you need? Or lacking a hill with a windmill—how about a small hydroelectric plant powered by a creek below the house that would divorce you permanently from dependence on power companies? Perhaps an underground or earth-contact house snuggled into the earth's cozy bosom during winter storms and summer heat, or a wood-burning stove (with an ample supply of firewood) to toast your toes on frosty nights, would be a better choice for you.

Most homeowners who are going to "natural" sources of energy employ more than one idea for reducing their dependence on outside sources. For example, the author's family installed a forced-draft wood-burning furnace and built a "solar wall" along a south-facing walk-out basement to help keep the wolf away from the door and the gas bills down.

The trouble is (seems like there's always a hitch) these "free" sources of energy are not free for practical use. For one thing, most equipment utilizing natural sources of energy has an initial high cost. You shell out a bundle in first cost to build energy-producing or energy-saving systems so you can have the pleasure of paying much lower or no energy bills in the future.

For another thing, not all natural energy sources fit everyone. Everyone cannot economically use wood for home heating—and dwellers in city condominiums may not be able to make much use of solar energy directly.

Wood-burning equipment today is more efficient and less obtrusive than stoves used 40 years ago.

For a third thing, many alternatives to petroleum energy are downright inconvenient. Take burning wood, for example. Many homeowners—particularly those in larger towns and cities—do not have access to a supply of firewood. Some of those who do have a wood supply close at hand may not be able to take advantage of this option because of ill health or physical handicaps.

Now, don't be discouraged by the previous three paragraphs. We merely want to point out that although there are bright opportunities in natural sources of energy, they all involve costs of some kind—either high capital costs of installation, inconvenience costs in operation, or both. As the old cracker-barrel economists put it: "There ain't no such thing as a free lunch."

However, that doesn't mean that a homeowner's only option to shivering in the dark is spending huge chunks of money all at once. Not at all. There are step-by-step solutions—compromises, if you prefer—that can let each individual mix and match those energy-producing and energy-saving systems which best fit his own situation. Part of any solution will be better, more efficient use of energy. The cheapest "source" of energy is conservation—whether energy is produced at home, bought off the pole or through the pipeline.

Consider this: a house with minimum insulation demands more heat energy to keep the temperature at 60 degrees in winter than does a well-insulated house (insulated to solar or electric heat specifications) with the thermostat set at a comfortable 72 degrees; an invest-ment in adequate insulation, weather-stripping, thermal-pane windows or shutters probably will pay off faster than about anything else; you'll need to stop heat thieves before you can realize full benefit from any system—conventional or natural energy.

You can convert your home to natural sources of energy without sacrificing much comfort or convenience. For example, you could buy two or three small wood heating stoves for different areas of your home, and spend all day and all night refilling fireboxes and fiddling with dampers to keep a comfortable temperature. Or, you could invest the same money (or less as you'd need only one flue) in a central wood-burning circulator or furnace that is thermostatically controlled and distributes heat evenly to all parts of the house.

Much the same is true of other natural energy systems. It costs little more (sometimes much less) to design and install a system or combination of systems which lets you enjoy both comfort and convenience. All the answers are not yet in on equipment, methods and applications for using natural energy for heat and power. But you can design and install natural energy systems on the basis of how well they pay a return on your investment within a reasonable time, and you don't have to disrupt the way your family lives to do it.

A solid-state, low-voltage controller can operate pumps, valves, dampers and blowers as efficiently and automatically as your wall thermostat controls your automatic furnace. Wind- and water-powered electric generation equipment can be controlled automatically. The point is, if you must stand by to manually operate equipment, you are not making much headway toward energy independence.

In the pages ahead, you'll find a great many ideas on natural energy systems and how homeowners are putting them to work: solar, wood, wind, water, geothermal, underground and earth-contact homes and many other energy sources. Some of these systems work better in certain situations or areas of the country; all of them cost money to build and install. Hopefully, you will find one or several ideas for making your own dwelling more energy efficient. That's the purpose of this book.

In Chapter 1, you'll find tips on how to save energy with little or no investment in money, simply by changing the way you operate the system you already have. Chapter 2 goes into retrofit or add-on systems that can be installed in existing homes at low cost. This section also describes how to figure the "pay-back" period on natural energy components.

In Chapter 3, we discuss components and equipment you'll want to consider when you buy or build a new home, or make a major modification of an existing house, to get the most energy value from each construction dollar spent.

Chapters 4 through 9, go into detail on specific aspects of natural energy: solar space and water heating, attached solar greenhouses, wood, geothermal, electric power generation and control systems for various installations.

In Chapter 10, you'll read about several homeowners around the country who have combined natural energy systems in their homes. Chapter 11, the "money" section, unravels the financial side of natural energy: tax incentives, loans, grants and resale value of energy-efficient homes.

In Chapter 12, you'll read about several U.S. communities that have instituted natural energy programs on a district-wide basis, with tips and advice from local residents on how to organize a community energy program.

Appendices list other sources of information on natural energy and energy conservation, a comprehensive directory of manufacturers and distributors of natural energy equipment.

Whatever you decide to incorporate in the way of "natural" energy systems—whether in an existing house or one now in the planning stage—count the costs and compute the "pay-out" for the systems or components you consider. If you live where the sun doesn't shine enough in winter to brew a good cup of tea, go slow on planning elaborate and costly solar installations.

If you're planning to convert to wood as a source of heat energy, consider the supply of firewood available—not only now, but in the future. Dependence on commercial wood cutters may not be any more comforting than dependence on the gas company.

Do some long and careful figuring if you plan to spend money on wind- or water-powered electrical generating plants. At the present time, dependable plants cost several thousand dollars for each kilowatt of generating capacity. If you capitalize the investment, for several years, even at current rates of interest, the interest on capital costs alone would buy a lot of power off the pole —even if commercial electric rates double.

So, hang onto those energy bills you paid this past year—at least until you get through Chapter 3. The money you now spend for energy is the best guide you have to decide how much you can afford to invest in natural sources of energy.

1
ARE YOU GETTING YOUR ENERGY DOLLAR'S WORTH NOW?

"As petroleum sources of energy become scarcer and more costly, our life styles will be simpler and more basic—but that's not necessarily all bad."

William Hughes
Electrical Engineer
Oklahoma State University

Do you still have those bills handy that show what you spent for heat and electricity over the past year?

Take another look at them now. Which months are high? Which are low? If you live in the northern United States, winter heating no doubt claims a large share of your energy dollar. If you live in a more southerly clime, your costs for heating and cooling may be split about 50-50.

Energy costs *will* go higher. In this age of uncertainty, that's a sure bet. The amount you spend per kilowatt hour of electricity, per gallon of fuel oil, per cubic foot of natural gas will be up in the next 12 months.

However, even if you live in a conventionally-heated house or apartment, those monthly bills are not necessarily fixed expenses—like rent or the mortgage payment. You probably won't have much success at trying to negotiate a lower electricity rate from your utility company. But you can negotiate a lower rate of energy *consumption* with your family.

The place to start is with an energy "audit" of your home. Your local electrical utility is required by law to perform an energy audit for its customers, either free or at a nominal charge. But you can learn a lot about your own pattern of energy consumption just by studying your energy bills. This will not hold down the price you pay for each *unit* of energy, of course, but it should point out some potential areas for energy conservation.

Because energy has been so cheap for so long, most of us have developed inefficient—even wasteful—habits of energy use. So what if I leave a 60-watt light burning when I leave the room? It would have to burn for 16½ hours to use up one kilowatt-hour of electricity, and that kilowatt-hour still costs only a nickel or so. Who cares if I let the hot water run all the time when I'm shaving? It costs only a few cents to heat the water. Big deal!

It *is* a big deal, when we multiply all those careless and wasteful habits we've acquired. If all the 200 million or so people in the U.S. left 60-watt bulbs burning for an unnecessary hour, that would total some 120,000 kilowatt-hours of electricity—enough to provide a whole year's electrical service for nine or 10 average homes.

But let's concentrate on *your* household, and the out-of-pocket energy expenses to run it. Unless you've already analyzed your energy consumption and what it costs, you may be surprised at how much energy you could potentially save just by operating your home more efficiently. And you can do it without sacrificing much comfort or convenience.

Most of the tips and techniques outlined in this chapter cost you nothing, or very little. They involve changes in your energy consumption habits.

Home Heating and Cooling are the big energy expenses for most families. In the average home, about $7 out of each $10 spent for energy goes for heating and cooling.

The heating zone map on page 12 shows how much of your winter heat bill you can save by turning down the thermostat by five or eight degrees from its usual setting. Let's say you live in Zone 2, and decide to turn your ther-

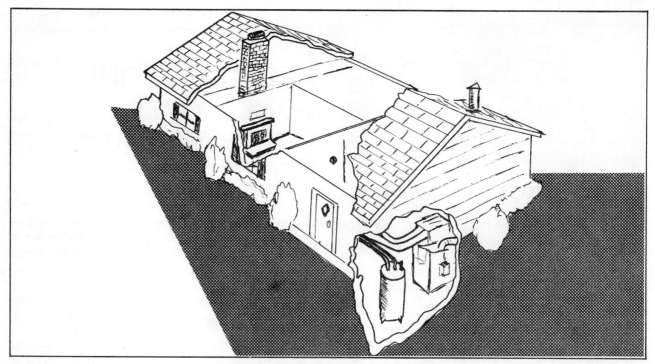

All of the areas indicated are potential heat thieves in the average single-family house.

mostat down by five degrees—from 72 degrees F to 67 degrees. How much will this save on January's heating costs?

If you heat with gas or oil, and use this fuel for no other purpose, merely multiply your January bill by $0.17. For example, if your January heating bill is $150, you can save a potential $25.50 for that month alone merely by keeping your thermostat five degrees under the usual setting. If you have electric heat that is not on a separate meter, you'll probably have to estimate how much of the total January electric bill is for heating—but the same percentage of savings applies.

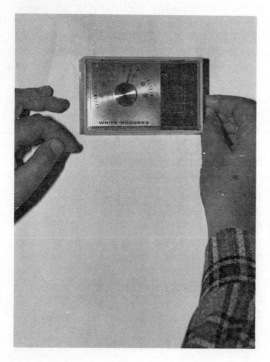

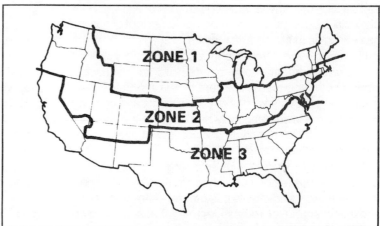

Turning the heating thermostat down by five to eight degrees can save considerably on heating bills. Find your heating zone on the map, then read the column below for that zone to see what you can save by dropping your thermostat setting:

	Zone 1	Zone 2	Zone 3
5° turn-down	14%	17%	25%
8° turn-down	19%	24%	35%

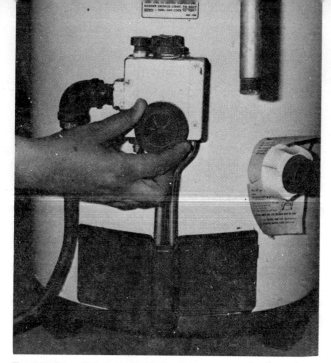

Turning down the water heater setting to 110-120 degrees can save $12 to $35 per year.

Much the same thing holds true for summer air conditioning. If you keep your air conditioning thermostat set at 72 degrees F., you can save about three percent of your energy bill for each degree you turn up the setting above 72 degrees F.

These changes, in the way you operate your home, cost you nothing. And, depending on where you live and what kind of heating-cooling system you have, the annual savings for turning down the thermostat in winter and turning it up in summer can save $35 to $120 on annual heating and cooling costs.

Here are some other cost-free ways to operate your home to save energy—and money:

1. Turn down the heat in rooms that are unoccupied for most of the day. Close heat registers, or turn down the thermostat on electric baseboard heaters. (If you're operating a heat pump, do not close vents and registers—this could damage the machine.)
2. Do not block radiators or forced-air registers with drapes or furniture.
3. Close fireplace dampers when the fireplace is not in use. An open damper on a 42 inch square fireplace can draft eight percent of your heat up the chimney.
4. At night, close shades and pull curtains across the windows. If you have single-pane windows without storm sashes, this can reduce the heat loss through the glass by 25 percent.
5. During winter days, "track" the sun with window shades and curtains. Open shades on: east windows in the morning, south during mid-day, and west in the afternoon, to let the sun warm the house interior. This is the simplest form of solar heating.
6. Place throw rugs against doors to stop drafts.

7. Keep doors and windows near the thermostat tightly closed. If the thermostat is located where drafts hit it, your heating system will work overtime.
8. Dust or vacuum radiator surfaces and heat register grills often.

Hot water is the second biggest user of energy in most homes, right behind space heating. If your household is typical, 15 to 20 percent of your total annual energy bill goes to heat domestic hot water. Here are some no-cost ways to save on energy needed to heat water:

1. Lower the heat setting on the water heater control to 110-120 degrees. (Some heaters have a dial that reads "high-medium-low"—set the control pointer at the low side of the medium setting). Water at 120 degrees is hot enough for all household uses, even if you have an automatic dishwasher.
2. When shaving, washing dishes, hands or faces, fill the sink with warm water, rather than let the water run.
3. After a bath, let the bathtub water sit until it cools to room temperature. Moisture carries heat better than dry air, and the humidity added to air in the house helps you feel comfortable at a lower temperature.
4. Use cold or merely warm water to wash those fabrics that do not demand hot water. Rinse clothes in cold water.
5. Flush sediment from the bottom of your water heater by drawing off two or three buckets of water from the heater drain. Do this every six months. Sediment and sludge can build up to insulate water from the heating element and lower the heater's efficiency.
6. If you'll be away for several days, turn off an electric water heater, or turn a gas heater's burner to "pilot." If your gas water heater has electric ignition, turn off the unit completely.

Here are some cost-free ways to save energy in the *kitchen*:

1. When possible, use pans with wide, flat bottoms that cover the stove burner. More heat goes to the pan; less is lost to surrounding air.
2. Keep range-top burners and reflectors clean; they reflect heat better.
3. When using the oven, try to cook more than one food at a time. Use a clock or the over timer; don't continually peek into the oven to check on baking progress. Each time you open the oven door, 25 to 50 degrees of baking heat is lost.
4. Don't use range burners or ovens to heat the kitchen.
5. Use cold water, rather than hot, to operate garbage disposers. This not only saves energy needed to heat water, it's also better on the appliance. Grease solid-

ifies in cold water and can be ground up and washed away.

6. Don't boil water in an open pan. Water boils faster and uses less energy in a kettle or pot with a lid.

7. If you use an electric range, get in the habit of turning off burners a few minutes before the elapsed cooking time. The burner element will stay hot long enough to finish the cooking without using more electricity.

8. Operate automatic dishwashers only with full loads. The average dishwasher uses 12 to 15 gallons of hot water per cycle, whether it is full of dishes or not. Let dishes air dry after the wash and rinse operations. If your machine does not have an automatic air-dry setting, turn the dishwasher off after the final rinse and prop the door open a little. Don't use the "rinse-and-hold" feature on your dishwasher—it costs you about five extra gallons of hot water.

9. Don't keep your refrigerator-freezer too cold. The fresh food compartment of the refrigerator should be set at about 38 degrees F.; the freezer section at 5 degrees F. Regularly defrost manually-defrosted models. Frost build-up increases the energy needed to keep the system running.

In the *laundry:*

1. Unless your washer has a small-load attachment or variable water level, fill (but don't overload) the machine for each washing.

2. Use the suds-saver feature if your machine has one. This lets you use one tubful of hot water for more than one load. Or, you can do the same thing by washing one load of clothes through the wash cycle, then removing them and washing a second load in the same suds. This saves both hot water and detergent.

3. Don't use too much detergent. Oversudsing makes your washer work harder, and can damage the machine.

4. An old-fashioned clothesline can save a lot of energy to dry clothes. You can install a clothes line for less than $25 and save that much the first year in costs of operating the electric clothes dryer.

5. When you do use a dryer, keep the lint screen and exhaust vent clean. Clogged screens and exhausts restrict the flow of air and cause the machine to use more energy.

Using a "solar clothes dryer" can save up to $50 per year, compared with using an electric dryer.

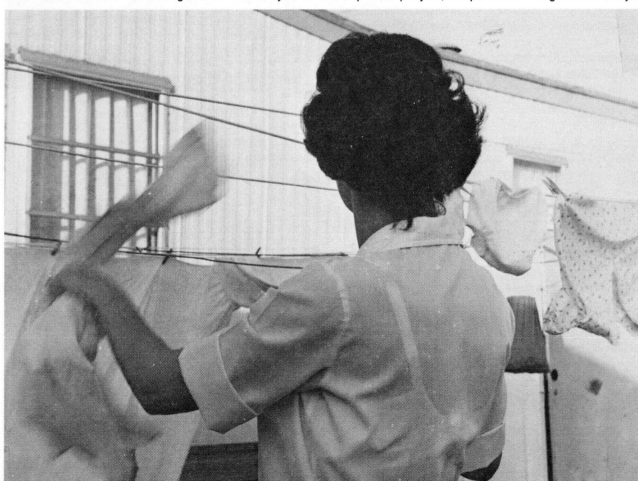

6. Dry clothes in consecutive loads to take advantage of the hot dryer drum. Save small, lightweight items to dry last—you may be able to dry them after you turn off the heat, with heat retained in the dryer from earlier loads.

7. Electric clothes dryers can be vented indoors in winter. Simply remove the exhaust pipe from the outdoor vent connection, cover the end of the pipe with a piece of cheesecloth or an old nylon stocking and direct the warm air into the house. Plug the outdoor vent to keep cold air out. Be sure to clean the cheesecloth or nylon often to maintain good air flow. Gas dryers should always be vented outside for safety. The sketch on this page shows how you can attach heat radiating "fins" to recover part of the heat exhausted through the ventilating pipe.

Home Lighting offers other no-cost opportunities to save energy. Here are a few, you probably can think of others:

1. Less than 10 percent of the total energy used in the average home goes for lighting, but most Americans overdo it. It's good energy conservation to cut back on the wattage and number of the lights you use.

2. Zone lighting is one way to save energy. Concentrate lighting in areas where reading or working is done, and where good lighting is needed for safety—stairways, for instance. Light intensity decreases with the *square* of the distance from the source. If you are reading this page with a light four feet away, you'd need four times as much light eight feet away to illuminate the page as well. Indirect lighting may be romantic and mood-setting, but it is not energy efficient.

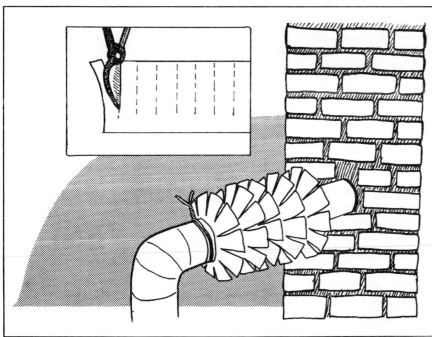

An electric clothes dryer can easily be vented indoors in winter, to save the heat generated. Gas and fuel oil burning dryers, water heaters and furnaces should always be vented outdoors, but light-metal heat radiating fins can be installed to trap some of the heat. Aluminum lawn edging can be cut as shown to make the heat-exchanger fins.

3. Use long-life incandescent bulbs only in hard-to-reach places. They use more energy than ordinary bulbs.
4. Keep lamps and lighting fixtures clean. Dirt absorbs light.

We've talked about ways to save energy in your home, without spending any money to do it, merely by changing the way you do things. How much can you save by following a conscientious energy-saving plan? It's hard to say, but perhaps 25 to 30 percent of the total energy bill, if you really work at it. If your annual energy bill is $1,000, that's $250 to $300 savings.

Now, let's take a look at some low-cost ways to save even more energy—and money. If changing the way you manage energy use in the home will save $250 per year, how much *more* can you save by investing that $250 in still more energy-saving devices and methods?

We'll start again with the major energy expense for most homeowners: heating and cooling the living space. The best, quickest paying investment most of us could make would be to increase the insulation in our homes and to weatherstrip doors and windows. We'll go into that more thoroughly in the next chapter. For the time being, let's stick with the little things; out-of-pocket expenses of less than $250 that can return big savings in energy costs.

1. Have your oil or gas furnace serviced each year. This can save you 10 percent on heating fuel consumption, and if you have it done in summer, many servicemen offer off season rates. While he's at it, have the serviceman check the combustion efficiency and firing rate of your furnace.
2. Clean or replace filters in forced draft heating systems once a month during the heating season; and on air conditioners each month during summer.
3. Install attic ventilators to pull superheated air out of attics.
4. Use "duct" tape to seal cracks and holes around doors and windows.
5. Install temporary storm windows made of 6 mil. thick plastic taped over the window. This is an especially good idea for renters. There are also storm window kits of rigid plastic on the market. One disadvantage of homemade storm windows is that they are difficult to remove if you want to open a window on an unusually warm winter day. Perhaps a better choice for you would be. . .
6. Shutters, made of rigid foam, Thermax or other insulating material. You can either cut the shutter to a size that will friction-fit snugly inside the window frame, or build a sliding track frame for the shutter.

Temporary storm windows can be built from kits or do-it-yourself materials available at most hardware stores. (Courtesy of Dept. Housing and Urban Development)

7. If you have a wood stove, furnace or free standing fireplace, you may want to install a heat reclaimer on the flue pipe. These attachments recover a good percentage of the heat that would otherwise escape up the flue. Many heat reclaimers have a small electric blower, and can be ducted short distances—to an adjoining room.
8. Install a whole house ventilating fan in the ceiling or an upstairs window, to pull cool air into the house. This can save a lot of energy, even if you use air conditioning during the hot part of the day. The fan keeps air moving through the house and keeps you fairly comfortable even up to 85 degrees, unless the relative humidity is high. A ceiling fan uses about 900 watts of electricity, compared with 9,000 to 10,000 for a central air conditioner.
9. If you have a fireplace, make sure the damper is in good working order (and closed when the fireplace is not in use). For older, open-masonry fireplaces, chimney-top dampers can be installed, with a control rod to allow the damper to be operated from inside the house. Also, you may want to install glass screens over the front of the fireplace, to cut down on the loss of heated air.

In most homes, there are opportunities to save energy required to heat water, by spending a little time and money. For example:

1. A hot water faucet that leaks just one drop per second can waste 60 gallons of hot water each week. Leaky

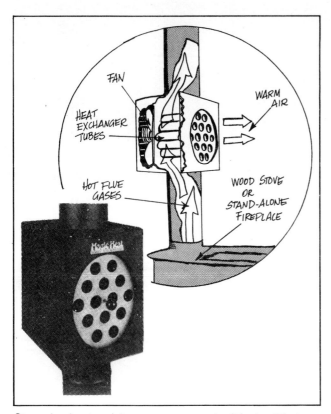

Stovepipe heat reclaimers recover much of the heat that would otherwise escape up the chimney. Some models feature a small electric fan to circulate heated air into the room.
(Courtesy of Magic Heat)

faucets—both hot and cold—should be repaired or replaced right away.

2. A water-saving shower head with a flow restriction valve is inexpensive and easy to install. You need only three to four gallons of water per minute for showering. Reducing the flow to this rate from about eight gallons per minute, and by turning off the water while soaping and shampooing, an average family of four can save more than $20 a year in water heating costs.

3. Install an aerator in the kitchen sink faucet. This device reduces the flow of water without noticeably restricting pressure.

You can spend money to save more money where home lighting is concerned. Here's how:

1. Use fluorescent lights wherever you can. A 25 watt fluorescent tube gives off as much light as a 100 watt incandescent bulb, but uses 75 percent less energy. Fluorescent lights are especially useful over kitchen counters and sinks, and to light bathrooms.

2. If you are replacing floor or table lamps, consider buying new ones with three-way switches. With these, you can keep light levels low when intense light is not needed, as when watching television, then turn the intensity up a notch or two for reading or sewing.

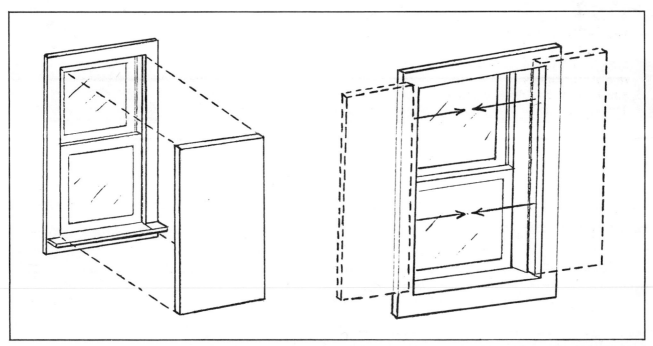

Insulating shutters can be cut to friction-fit individual window opening, or built as an insulation-and-wood "sandwich" used as sliding doors to cover windows at night. (Courtesy of University of Missouri)

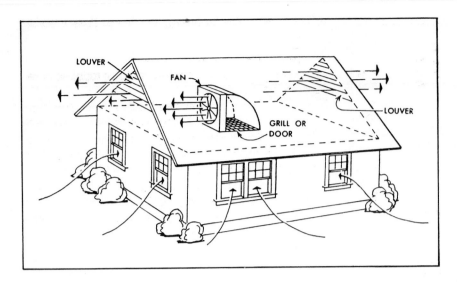

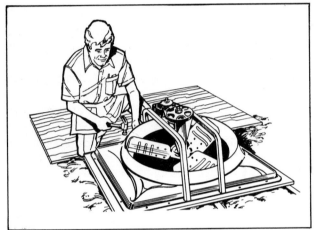

A centrally-located attic fan can help lower summer air-conditioning costs. Fans can be installed quickly in the ceiling, to exhaust heated air from the attic and pull cooler outside air into the house. (Courtesy of USDA and Kool-O-Matic)

Shower heads with volume control and shut-off valve can save gallons of hot water over a year's time (Courtesy of Harcraft)

3. Look into light dimmer switches that let you adjust the intensity of room lighting. We'll have more to say about them in Chapter 3.
4. Use lower wattage directional floodlights for outdoor lighting. A 50 watt reflector floodlight provides as much light, but uses half the wattage, as a standard 100 watt bulb.
5. In multi-light fixtures where all the light is not needed, use lower wattage bulbs, or leave some of the bulbs out. For safety's sake, you can use a burned-out bulb in the socket.

In this chapter, we've outlined just a few ways to save on energy costs by spending no money at all—or just a little. The efficient use of energy is necessary basic training for anyone who plans to install alternate, renewable sources of heat and power.

Most homeowners no doubt can think of other ways to make their own households more efficient in the use of energy. It's mostly an attitude. Instead of thinking: "I'm consuming only a few cents by letting this light burn"; an energy efficient householder gets into the habit of thinking: "I'm saving a few cents by turning this light off."

2
EASE YOUR WAY INTO NATURAL ENERGY

"More energy can be saved in the average home by overall insulation techniques than can ever be saved by the most expensive addition of solar hardware."

Gordon L. Moore
Mechanical Engineer
University of Missouri

This year, about eight million Americans will move into new housing units according to estimates by the nation's housing industry.

That's about two million chances to incorporate sound energy systems into the design of those homes, condominiums and apartments—from the ground up. Of course, whether a very big percentage of them *are* built to be energy-efficient is another matter.

But the real opportunity to save money on energy accrues to the rest of us; the more than 200 million who still will be living in homes and apartments somewhat older. For every family buying or building a new home this year, more than 50 families will continue to live in housing that existed at the beginning of the year—a few of us in housing that stood at the beginning of the century.

In the days of adequate fuel supplies (and today as well) excessive heat losses through poorly insulated ceilings, walls and large glass areas were "taken care of" by installing larger heating plants. A good many of us are stuck with such places, and most of us will have to make do with our present house—at least for the time being.

Admittedly, designing a natural energy home would be a more pleasant adventure if you could start with a clean sheet of paper and a bare building site. Many owners of older homes are making investments now that will give them a head start in surviving the years ahead where most experts see scarcer and more expensive energy. There are a good many ways to "retrofit" or add natural energy systems to an existing house; even for those that

were built before the world's oil spigot dwindled from a gush to a drip.

Some may have the money to build one of those "glamour" energy-producing systems we read about: an active solar heating set-up, or a huge windmill to crank out watts. They may be able to afford extensive remodeling and rebuilding to update the insulation, electrical system and other energy-related features in their homes. That's a move that will pay bigger and bigger dividends in years to come.

Others, however, less affluent or on tentative terms with the banker, will have to ease their way more gradually into natural energy systems. Part of that package, as we mentioned in Chapter 1, must be conservation—more efficient use of the energy that is consumed.

If homeowners are going to escape the tyranny of higher and higher oil costs, they will need to cut back on energy use while they are moving to natural energy systems: wood, sun, wind, water and earth. For most of us, there are still going to be bills to pay. The first step to trimming those energy bills is to learn to live comfortably on less energy; by practicing those energy saving habits outlined in the previous chapter and by investing in those home energy systems that will give the greatest return for each dollar invested.

The next move is to evaluate alternative energy systems on the same basis—return on investment—for your own situation. Wood heat is the most readily available retrofit system for many existing homes. Trees are efficient collectors and storage of solar energy, and much of the U.S. has surplus wood that is accessible to a great many people.

Add-on solar systems are another possibility as an auxiliary source of heating energy. Air heaters, to provide at least a portion of the needed winter space heat, built along south facing walls can be constructed of relatively inexpensive materials. Solar water heaters can take much of the load off conventional water heaters—on a year 'round basis.

We'll talk more about retrofit systems, but right now take still another look at your energy bills for the past year. Multiply the total amount you spent times 10 (for an average interest rate of 10 percent) and capitalize your energy dollars to see what kind of investment you could justify on the basis of present energy costs.

Let's assume that you will invest the money in a system with a life expectancy of 10 years or more. And we'll say, for example, that you spent $1,000 total last year for heat and power. If you paid for it in cash, you could afford to spend $10,000 for alternate energy systems that would provide *all* your heat and power needs for the next 10 years. The $10,000, spread out over 10 years, would come to $1,000 per year. There are tax incentives that let Uncle Sam pay part of the cost of alternate energy systems—we'll go into that subject thoroughly in a later chapter. And, if you must borrow the money, interest costs extend the time the system takes to pay for itself. On the other hand, if conventional energy costs double in the next 10 years, such a system would pay for itself much sooner. Each homeowner will need to compute these features as they apply to his situation.

INVEST IN HOME HEATING (AND COOLING) FIRST

For most homeowners, investing in equipment to produce *all* the heat and power needed is somewhat less than practical. Most of us need to put priorities on limited energy dollars. The wisest course of action is to invest first in materials, systems and equipment which will give the greatest rate of return. In most homes, for example, space heating and domestic water heating demand attention first.

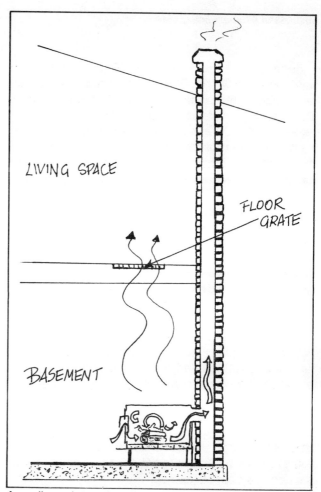

A small wood stove installed in the basement can supply a great deal of heat to rooms above.

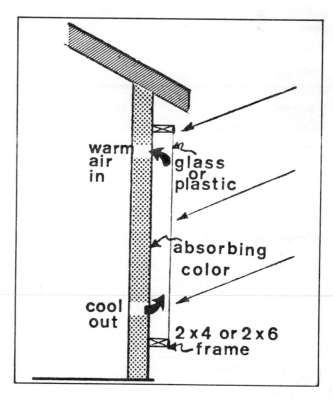

Solar installations need not be elaborate and expensive to take some of the home heating load. (Courtesy of University of Missouri)

Suppose you could spend $4,000 on material and equipment that would cut your annual energy bill in half; install more insulation and build a solar collection system. That means your annual energy bill would be only $500, rather than $1,000, in terms of current dollars (not counting what inflation does to the dollar in years to come). The $500 you save amounts to a 12½ percent annual return on your $4,000 investment, again in terms of today's dollars.

The beauty of this kind of capital investment is that it keeps on paying, even after you have recovered your initial outlay. Insulation will continue to stop heat and energy losses for as long as the house stands. If a solar heating system is built of durable materials (as it should be) there isn't much that can go wrong with it in 10 years—or 30 years.

But, before any *natural* heating system is considered, the homeowner should see to it that walls, ceilings, floors, basements, crawl spaces—in short, *all* exposed surfaces—are well insulated. More energy (and money) can be saved in a typical home by insulating and weather-stripping than by virtually any other investment.

So, if your decision—forced by limited dollars—comes down to an "either-or" choice between spending $1,000 on a solar energy collector or spending $1,000 to bring your home's insulation up to par, spend the money for insulation. It will pay off quicker.

The one big unknown in figuring the pay-back on energy investments is this: how much will conventional energy costs increase, and how fast? OPEC (The Organization of Oil Exporting Countries) estimates that petroleum costs will rise by an average of 7 percent through 1985. When you add in a factor for inflation in the U.S., the dollar cost of energy could climb 12 to 14 percent or more each year.

If the average life of the equipment listed above is 15 years and the OPEC projections for petroleum costs prove to be accurate, all of these systems except wind generated power would repay the initial cost in less than 10 years.

INSULATE

As you can see by the chart below, a dollar invested in insulation and weather-stripping, returns a potential saving of 15 percent per year in energy costs. We will not go into great detail on insulation in this book. There are excellent guides on home insulation listed in Appendix II. Even if your energy-saving strategy includes installing a wood stove, solar heater or other "natural" heating system, the investment in adequate insulation will let these

Plan to invest money first in those areas with the biggest pay-off, which can help you get a lot more for your investment.

Before	After	Cost*	Yearly Saving**	Return on Investment
Conventional heating (Gas, oil, electric)	Adequate insulation; weather-stripping; storm windows	$1,000	$150	15%
Conventional heating	Wood stove (cut your own wood)	$1,500	$200	13%
Conventional heating	Passive solar system	$2,000	$200	10%
Electric or gas water heater	Pre-heater in solar or wood system	$1,500	$120	8%
Commercial power	Wind generated power to provide 50% of electricity needs	$6,500	$300	4½%

*Initial costs only—does not include maintenance and operating costs, if any.
**Based on a family of four living in a 1,500-square-foot single family house in the Midwest.

systems operate more efficiently. An early step in planning any natural heating system is to check on the insulation of your home. Adequate ceiling and wall insulation,
caulking, weather-stripping and other energy conserving
devices cut the heating and cooling load for whatever
system you use.

Start by checking how much insulation your home now
has, and what kind. Check first in the attic. Heat rises.
Your ceiling should have at least 6 inches of mineral wool
or the equivalent. Identifying the type and thickness of
wall insulation can be more difficult, but you may be able
to remove a switch plate or electrical outlet plate to examine the insulation behind it.

The map on this page shows recommended "R-values" for different regions of the U.S. The R-value of
any material refers to its resistance to heat flow. Federal
regulations that went into effect in November, 1979, require that all insulation materials be labeled with an
R-value for any given thickness.

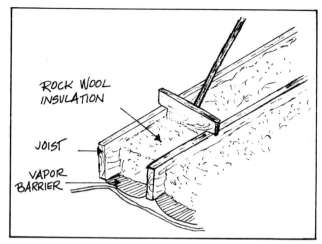

Heat rises—the first priority for insulation is the attic. Poured mineral wood insulation can be added over existing insulation.

Recommended insulation for various zones of the U.S. (Courtesy of U.S. Department of Energy)

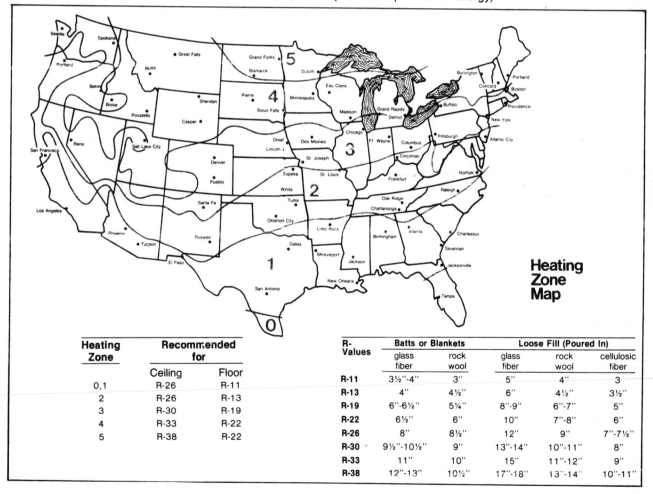

Heating Zone	Recommended for	
	Ceiling	Floor
0,1	R-26	R-11
2	R-26	R-13
3	R-30	R-19
4	R-33	R-22
5	R-38	R-22

R-Values	Batts or Blankets		Loose Fill (Poured In)		
	glass fiber	rock wool	glass fiber	rock wool	cellulosic fiber
R-11	3½"-4"	3"	5"	4"	3
R-13	4"	4½"	6"	4½"	3½"
R-19	6"-6½"	5¼"	8"-9"	6"-7"	5"
R-22	6½"	6"	10"	7"-8"	6"
R-26	8"	8½"	12"	9"	7"-7½"
R-30	9½"-10½"	9"	13"-14"	10"-11"	8"
R-33	11"	10"	15"	11"-12"	9"
R-38	12"-13"	10½"	17"-18"	13"-14"	10"-11"

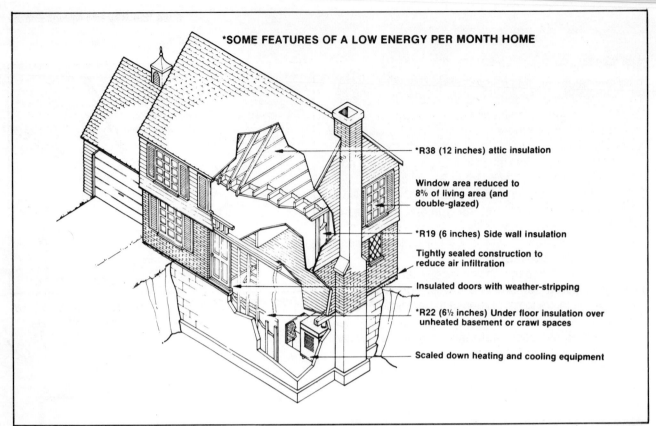

*R38 (12 inches) attic insulation

Window area reduced to
8% of living area (and
double-glazed)

*R19 (6 inches) Side wall insulation

Tightly sealed construction to
reduce air infiltration

Insulated doors with weather-stripping

*R22 (6½ inches) Under floor insulation over
unheated basement or crawl spaces

Scaled down heating and cooling equipment

Insulation is a key to efficient home heating and cooling, regardless of the source of energy.
(Courtesy of Owens-Corning Fiberglas)

Figure the total R-value of a surface (ceiling, wall, etc.) by adding the resistance of all materials. For example, normal ceiling materials (sheet rock, tile, etc.) have an R-value of about 2.0. Typical wall materials have an R-value of about 2.5. If your ceiling is of normal tile and has 4 inches of mineral wool in the attic above, the total R-value is approximately 18.

Here are R-values per *inch of thickness* for commonly used insulation and building materials:

Material	R-Value per Inch
Fiberglass; mineral wool batts	3.7
Fiberglass; mineral wool fill	4.0
Wood fiber	4.0
Vermiculite	2.27
Shredded pulp; paper	4.16
Expanded polystyrene	4.0
Expanded polyurethane foam	6.25
Ureaformaldehyde foam	5.00
Insulated sheathing board	2.63
Plywood	1.26
Fir or pine boards	1.3
Cinder block	.36
Concrete block	.24
Poured concrete	.08

Generally, the higher the R-value, the better the insulation. However, ease of installation, cost and the ability of a material to resist fire and moisture will influence your choice.

Unless you have need of a special material, select insulation on the basis of its cost per *net square foot* for a given R-value. Suppose you want to install insulation in your attic with a total R-value of 19. You price one kind of insulation that has an R-value of 3.2 per inch of thickness and costs 7 cents per square foot at the 1-inch thickness. Another type has an R-value of 3.7 per inch of thickness, but costs 8 cents per square foot. Which is the better buy?

You'd need 5.9 inches of thickness of the first type to total R-19. At 7 cents per square foot (for a 1-inch thickness), that's 41.3 cents per square foot when you have insulated to an R-value of 19. With the second type of insulation, at 8 cents per square foot, you'd only need 5.1 inches of thickness, at a cost of 40.8 cents per square foot to get R-19.

Getting more insulation over your ceiling may not be much of a problem. You can use insulation materials in batts, rolls, or loose fill to increase the R-value in the attic. Same thing is true of most basements and crawl spaces, where you'll be insulating the floor above. Batts or roll insulation can be stapled between floor joists.

Adding insulation to finished walls can be more trouble. You can hire a contractor to blow loose-fill insulation into wall cavities, or to inject foam into the walls. Or, if your dwelling is due for new siding, it's a simple matter to fasten rigid boards over the studs, then install the siding directly over the insulation.

A note on hiring insulation contractors: before you sign on the dotted line, check his reliability with your local Home Builders' Association and the Better Business Bureau. Also, get a warranty in writing to cover both work and materials. Be sure that vapor barriers, adequate ventilation and fire safety precautions are spelled out in the contract.

Insulation Materials

Here are some things you'll want to consider before you choose the insulation materials to add to your home:

Fiberglass or mineral wool batts and blankets are preformed sections and come with or without vapor barrier backing. They are fire resistant, and probably are the best choice for most do-it-yourself installation—particularly for attics and under floors. These materials are relatively inexpensive and readily available at building supply stores.

Caution: Fiberglass can cause skin and eye irritation.

Areas that should be insulated are (1) exterior walls, walls between heated and unheated parts of the house; (2) ceilings; (3) knee walls when attic is unfurnished; (4) between collar beams and rafters; (5) around the perimeter of slab; (6) floors over crawl spaces; (7) floors above unheated spaces; (8) basement walls; (9) behind rim or header joists. (Courtesy of Mineral Wool Insulation Assn.)

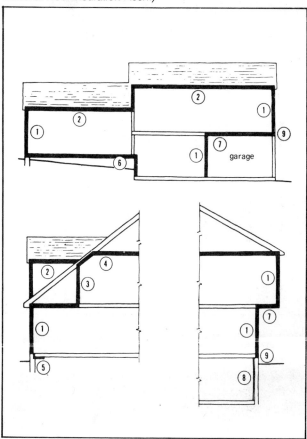

Crawl spaces under floors should be insulated. Batt or blanket type insulation is convenient to use in these areas. A vapor barrier of polyethylene film over the soil in the crawl space keeps insulation dry.

Water heater insulating kits are available, or you can add roll-type fiberglass insulation to the water heater tank and pipes. Insulate the hot water pipe coming from the water heater for at least three feet.

Contractor applied foams—urethane, ureaformaldehyde or polystyrene—are forced into place under pressure as a liquid, then harden to form insulation. They have a high R-value per inch of thickness and are often the only solution for adding insulation inside finished walls.

Caution: They are not fire-resistant. Ureaformaldehyde can cause bad odor problems if not properly mixed or cured. Foams should not be used where humidity is a problem, or where they will be exposed to direct sunlight.

Rigid plastic foams and glass fiber boards have a high R-value and are especially useful for basement walls or for outer sheathing between studs and siding.

Caution: Some rigid insulation is not fire resistant. All are relatively expensive.

Loose fill (pour-type) fiberglass, mineral wool and cellulose fiber products are easy to install in unfinished attics. These materials also can be blown into walls; however, they tend to settle over time.

Caution: Loose fill insulation requires a separate vapor barrier.

While you're insulating, don't forget hot water lines and warm air ducts that run through unheated crawl spaces and basements. Sheet-metal heating ducts and copper water pipes are the more common types and both can radiate a lot of heat to cool air.

Duct insulation can be bought at most building supply stores, in one- and two-inch thick bankets. For hot water pipes, you can use roll type fiberglass or preformed insulating sleeves. Insulating ducts and hot water pipes can save from $20 to $140 per year in energy costs; more if the same ducts carry cool air from an air-conditioning system in summer.

Insulating a normal sized water heater can save another $24 to $66 each year—more than enough to pay the cost of the insulation two or three times over. Insulating kits are made for both gas and electric water heaters, or you can merely wrap the heater tank shell with fiberglass and tape it in place. With gas water heaters, make sure that the air intake at the bottom of the tank or the exhaust vent at the top of the heater is not blocked by insulation.

DRAFT-PROOF WINDOWS AND DOORS

Windows are a mixed blessing. They are great for admitting light, providing a view out-of-doors and collecting some solar energy during the day, but are not good insulators against nighttime heat losses. In fact, single-pane glass acts as a heat exchanger to the outdoors, allowing about 10 times as much heat loss per square foot as a well insulated wall.

1

Before applying caulking compound, clean area of paint build-up, dirt, or deteriorated caulk with solvent and putty knife or large screwdriver.

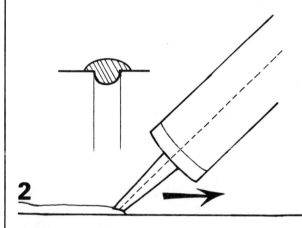

2

Drawing a good bead of caulk will take a little practice. First attempts may be a bit messy. Make sure the bead overlaps both sides for a tight seal.

A wide bead may be necessary to make sure caulk adheres to both sides.

4

Fill extra wide cracks like those at the sills (where the house meets the foundation) with oakum, glass fiber insulation strips, etc.)

FOUNDATION SILL

5

In places where you can't quite fill the gaps, finish the job with caulk.

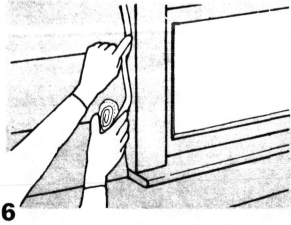

6

Caulking compound also comes in rope form. Unwind it and force it into cracks with your fingers. You can fill extra long cracks easily this way.

Source: Dept. of Housing and Urban Development.

WEATHERSTRIP YOUR DOORS

AN EASY DO-IT-YOURSELF PROJECT

You can weatherstrip your doors even if you're not an experienced handyman. There are several types of weatherstripping for doors, each with its own level of effectiveness, durability and degree of installation difficulty. Select among the options given the one you feel is best for you. The installations are the same for the two sides and top of a door, with a different, more durable one for the threshold.

The Alternative Methods and Materials

1. Adhesive backed foam:

Tools

Knife or shears,
Tape measure

TOP VIEW

Evaluation — extremely easy to install, invisible when installed, not very durable, more effective on doors than windows.

Installation — stick foam to inside face of jamb.

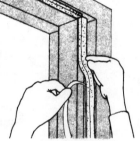

2. Rolled vinyl with aluminum channel backing:

Tools

Hammer, nails,
Tin snips
Tape measure

TOP VIEW

Evaluation — easy to install, visible when installed, durable.

Installation — nail strip snugly against door on the casing

3. Foam rubber with wood backing:

Tools

Hammer, nails,
Hand saw,
Tape measure

TOP VIEW

Evaluation — easy to install, visible when installed, not very durable.

Installation — nail strip snugly against the closed door. Space nails 8 to 12 inches apart.

4. Spring metal:

Tools

Tin snips
Hammer, nails,
Tape measure

TOP VIEW

Evaluation — easy to install, invisible when installed, extremely durable.

Installation — cut to length and tack in place. Lift outer edge of strip with screwdriver after tacking, for better seal.

So-called insulating glass—two panes with an air space between—loses only half as much heat as a single pane. But this is still about 5 times as much as an insulated wall loses.

Combination storm and screen sashes are convenient, but cost $30 or more per window to install. The air space between glass (or plastic) panes is what provides the insulation. You can build and install very effective storm windows yourself from aluminum channel framing and glass bought from the lumberyard; or you can buy transparent plastic storm window kits. Either route will cost about $10 to $15 per window, but, unless you modify the window frame to allow easy installation and removal, homemade storm windows can be more trouble than custom ones when you need to open a window.

You can save as much as 15 percent per year on your energy bills by adding storm windows and storm doors, provided that you also seal drafts around these openings.

Test doors and windows for air tightness by moving a lighted candle around the frames and sashes of closed windows and doors. If the flame flickers, you need weatherstripping and caulking. Also check the bottoms of exterior doors. If you can slip a quarter under the door, it needs weatherstripping. Caulking and weatherstripping are not big projects for most homes. You can do it yourself, with a minimal cost for materials.

Caulking compounds for sealing stationary joints around exterior door and window frames come in a variety of materials and prices. Oil or resin base caulking is low cost, but is least durable when exposed to weather. Latex, butyl and polyvinyl types are somewhat more expensive, but last longer. Silicone and polysulfide base compounds are most expensive, but also most durable—some experts say these are the best buy in the long run.

Caution: Don't use lead-base caulk; it's toxic.

You'll need some form of weatherstripping to seal the moving joints at the perimeter of exterior doors and windows. Felt or foam rubber strips are relatively inexpensive and easy to install—but not very long lasting. Rolled vinyl with metal backing strips is medium priced and easy to install. It lasts longer than felt or foam, but is visible when installed. Thin metal spring strips are fairly expensive and tricky to install, but they are an invisible, permanent seal when properly installed. Most expensive—and most difficult to install—are interlocking metal channels, but they are an excellent weather seal.

OTHER ENERGY SAVERS

The tightest door doesn't save much heat when it's open. One way to cut down on the heat that rushes out every time you open the door is to add a vestibule, creating an air-lock. Those entry locks designed either into the interior of the house or added to the exterior, reduce energy losses by providing two doors, only one of which is normally open at any time. Perhaps your house has an open porch or stoop area that can be enclosed to form an entry lock. If the door opens into a long hall, you can frame up a cross wall with a door frame. Such an arrangement also saves on energy to cool a house in summer.

A vestibule, or entry air-lock, can be added to most homes, either outside or inside the present exterior door.

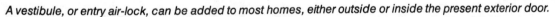

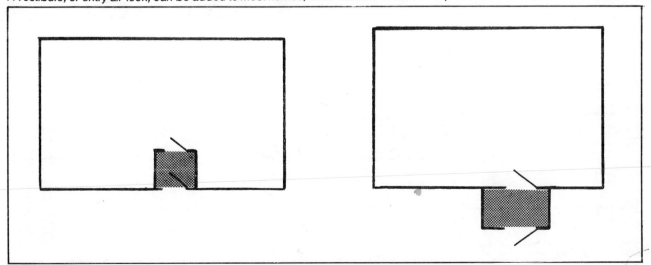

Insulating window shades are another good winter-summer idea for saving energy. Several companies make them now; some in retractable pull-blind styles.

Or, how about awnings to shade windows in summer? Your grandparents probably used them on south and west windows before home air-conditioning became popular. New synthetic canvas materials are durable and color-fast.

A clock thermostat (a so-called "automatic set back" furnace controller) that automatically adjusts home temperatures for normal family activities can pay for itself in a couple of ways. You can set the thermostat to drop a given number of degrees at bedtime; the unit automatically moves the room temperature up again by early morning. Or, in homes where everyone is away for most of the day, the controller can be set to turn up the heat by 5:00 or 6:00 P.M. when everyone comes home.

These thermostats cost in the $100 to $150 range, and can control air-conditioning systems as well as heating plants.

Even in winter, attics can collect a lot of heat on sunny days. A thermostatically-controlled attic fan can be ducted to blow the heated attic air into living spaces or into a basement or crawl space.

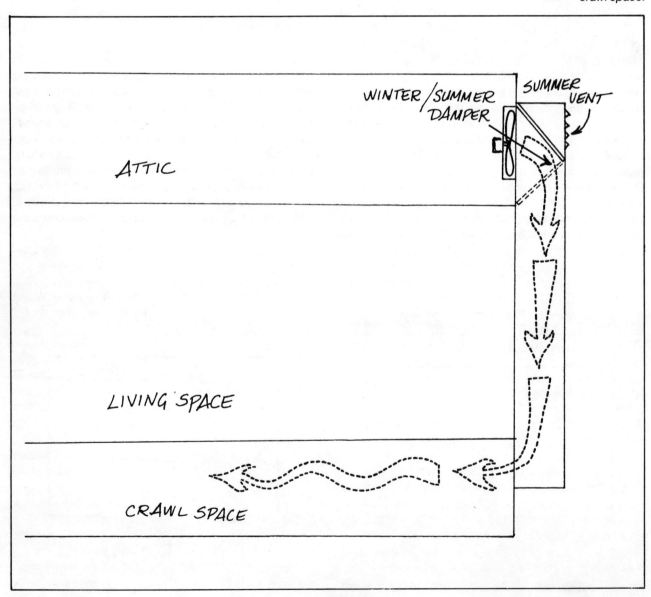

SUPPLEMENTAL HEAT FROM NATURAL SOURCES

Now, having insulated and tightened up the old homestead—with the result of dollars saved and living spaces made more comfortable—you may be interested in adding a natural heating system to take up part of the load. We'll go into such heating systems as solar, wood and geothermal in chapters ahead for those homeowners who want to design more elaborate systems.

But for right now, we'll talk about using sources of energy to *complement* the system you already have installed and operating.

Wood is one of the better sources of supplemental heat for several reasons:

1. In most regions of the country, wood is available nearby at a fairly reasonable cost.
2. A wood stove and insulated metal flue can be installed rather easily, and without a terrific cash outlay, for a stove or wood-burning furnace wherever it's needed.
3. Wood is relatively easy on the environment.
4. Best of all, wood is a renewable energy resource.

Some 500,000 wood-burning devices will be installed this year in American homes. With about 400 different companies building stoves, fireplaces and furnaces to burn wood, it can be confusing to decide which wood-burner is for you. But it doesn't have to be all that complicated. You'll want to take into account the size and shape of your home, your existing heating system and how much you will depend upon wood to heat your home.

If your goal is heating efficiency, forget about fireplaces. An open fire adds a measure of cheeriness to the home, but at a price. Even the better heat-circulating models are only about 25 percent efficient, compared with 50 to 60 percent efficiency for a well designed stove or furnace. In fact, if your home has an old-style masonry fireplace, you may want to fit it with a flue kit and mount an airtight, thermostatically controlled wood stove on the hearth. You'll get a lot more heat out of your firewood that way.

Ben Franklin didn't know all there was to know about wood stoves! Modern airtight woodburners are designed for heating efficiency. The amount of heat you get from any stove will depend on what kind of wood you burn and, of course, how you operate the equipment.

To be technical about it, the wood itself doesn't burn. As wood is heated, gases and charcoal are produced and it is these that actually burn and yield heat. The more of these products that are burned, the higher the combustion efficiency of a wood-burning unit. Well designed wood stoves keep the wood at a temperature high enough to produce gases, and draw a controlled amount of air through pre-heating chambers, or a system of baffles. In these stoves, the hot gases pass through a second chamber or over a baffle, above the actual fire box, so that more heat is extracted from them before they exhaust up the flue.

Many airtight stoves are equipped with a bimetal thermostat to regulate the draft damper. Better controls do a remarkably good job of holding the heat at an even preset temperature.

Wood stoves are of two general types: radiant heaters and circulators. A radiant heating stove warms the surfaces in the room where it is located—by radiating heat to walls, floors, ceilings and furniture. A circulating heater has a chamber between an outer cabinet and the firebox, through which air circulates and is heated. Some have electric blowers to move more air through this heating chamber.

Quality wood stoves are typically made of either cast iron or steel plate. Both are durable materials. More important than the material, probably, is the construction, design and weight of the unit. Heavier metal is not as likely to warp and buckle, holds heat longer—and costs more.

A good airtight stove will cost more, but you get what you pay for. If you don't have a useable chimney now, plan to spend $1,000 to $1,500 for the entire installation.

Since the Arab oil boycott of 1973 and the subsequent attention on petroleum, wood stove manufacturing companies have sprung up like mushrooms. Some of these newcomers build excellent stoves; some turn out poorly made, flimsy stoves. Your best bet is to buy a stove produced by a reliable, established firm and handled by a competent dealer—even if it takes shopping to find what you want.

Does the stove or furnace have the Underwriters Laboratories (UL) seal? What kind of warranty does the unit carry? Better made stoves are guaranteed for at least 10 years. One manufacturer, New Hampshire Wood Stoves, Inc., warrantees their "Home Warmer" model for the life of the original buyer.

How much can you save by installing a wood stove? That depends on your winter heating needs, what kind of stove you buy, where you put it and how you use it. Francis Bessette, who lives in Grafton, North Dakota—where winters get really cold—figures he has saved $1,500 worth of fuel oil the past two winters by using an Ashley airtight wood heater.

You may not save that much by going to wood for part of your winter heating, unless you live near Bessette's latitude. But if you have an accessible supply of firewood, your money for a wood stove will likely be well spent. A quality wood burner should give trouble-free service for a great many years, and its value will increase as conventional forms of energy become more costly.

Another approach to lowering your heating bills is to retrofit your present home to make use of the energy of the

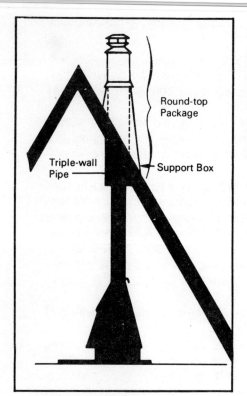

A-frame roofs pose no installation problems with the proper chimney package.

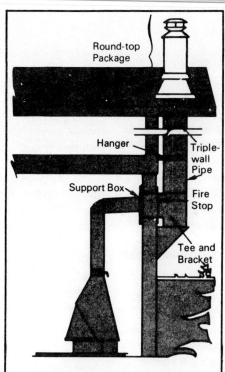

No need to cut through the attic. It's easy to elbow through an outside wall and vent straight up, using Tee and Bracket.

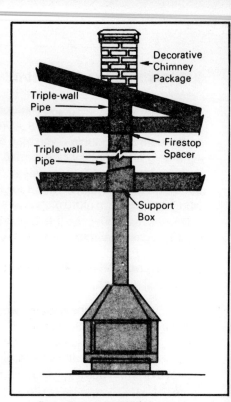

First floor installation in a two-story house. Chimney can be concealed in a second story closet.

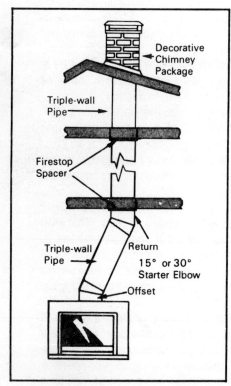

Preway chimneys clear upstairs obstructions with 15° or 30° elbows (all elbow kits include offset and return).

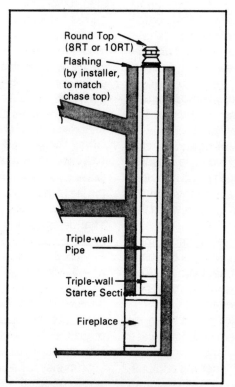

Today's space saving chase installations are a natural with Preway built-in fireplaces and chimney system.

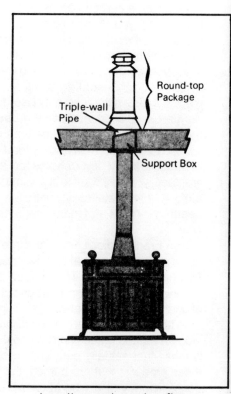

Installation through a flat roof is simple and the most economical of all.

Both fireplaces and wood burning stoves can be installed in a variety of ways: The six illustrations here detail some typical methods of chimney installation. (Courtesy of Preway)

Old-style fireplaces can be converted to more efficient heating by installing a damper panel and wood burning stove. (Courtesy of National Stove Works, Inc.)

sun for space or domestic water heating. Depending upon where you live, and the size and shape of your house, a solar collector can supply a good part of your winter heat.

However, here are some questions you should answer before you start building and installing collectors:

1. What kind of climate do you have? Solar space heating pays off better where winters are long but mostly sunny, and utility rates are high (which is about everywhere, these days).
2. Does your roof (or a long wall) have enough southern exposure to accommodate collectors? If not, do you have enough yard space to build a detached collector? Is the area shaded for part of the day?
3. Is the roof strong enough to support a large solar collector assembly without extensive shoring up?
4. Is your present heating system compatible with solar? Forced air heating systems are good companions for solar heated air systems—they operate at similar temperatures. Likewise, solar water systems can be retrofitted to some hot water home heating plants. Electric resistance heating is not as compatible, be-

cause the solar system needs an air or hot water distribution system.

J.W. Fish, of El Reno, Oklahoma, built a 344-square-foot collector (about 38 square meters) on the roof of his older brick home, and built an addition onto his house to hold the 3.5 tons of rock that stores the heat. Fish figures he has a cost of $2,200 in the entire system that will provide 40 to 50 percent of his winter heating.

The collector panels were framed with 2x6 lumber, and covered with 4x8 sheets of solar fiberglass. Second-hand aluminum offset printing plates, with two coats of flat black paint, were used to make the back of the collectors.

"We're using a 700 CFM (cubic feet per minute) blower to move heated air from the collector to the storage," Fish said. "Heated air for space heating is ducted either directly from the collector or from the rock storage into the house through the ceiling."

Cold air is returned through the crawl space under the house. Fish also ran two loops of water pipe through the storage compartment as a water pre-heater.

If you prefer to make a more subtle entry into solar heating, you may want to start with solar panels along a south wall. Ideally, a solar collector should be at nearly right an-

A simple solar retrofit system can be built of inexpensive materials, to collect and circulate solar heat into the crawl space. The heat migrates through the floor to the living spaces above.

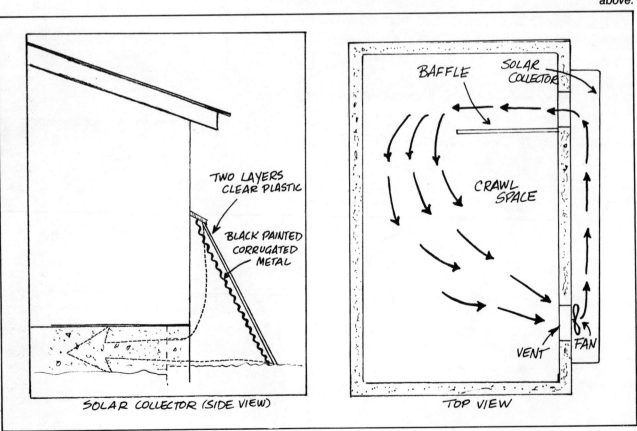

SOLAR COLLECTOR (SIDE VIEW)

TWO LAYERS CLEAR PLASTIC

BLACK PAINTED CORRUGATED METAL

TOP VIEW

BAFFLE   SOLAR COLLECTOR

CRAWL SPACE

VENT   FAN

gles to the sun's rays, but a vertical surface can pick up a lot of heat.

Several types of solar heaters—for both air and liquid—are available commercially. But all a solar air heater consists of is a glass or plastic covered box with insulation in the back. A solar water heat collector works on the same principle, but has water or some other fluid circulating through the collector to absorb and carry away heat. Any homeowner with average building skills can construct a collector to fit the space he has available.

Or, you may want to retrofit for domestic hot water only. A hot water system is generally simpler and less costly to build than a space heating set-up. There will be some plumbing and wiring involved, and perhaps circulating pumps to install. But a hot water system has the advan-

tage of being in use year 'round, and is therefore more cost effective than space heating in most of the U.S.

Okay, all this will cost money—how much can you expect to get back? Adding insulation, weather-stripping, caulking, storm windows and insulating panels should save about 25 percent of your winter heating bills. If you do the work yourself, you should get these projects completed for less than $1,000, for the average sized home. You have looked at your energy bills a couple of times here lately—how long will it take 25 percent of your heating bill to total $1,000? If you have a good idea of what percentage of your total heating needs the natural system will provide, you can use the same method to compute the "pay-back" on a wood stove, solar collector or other installation.

Solar collector panels built onto the roof of J. W. Fish's house have little of the "afterthought" appearance of some solar retrofit projects. Fish also gave some thought to appearance when he built the heat-storage bin of the same brick pattern as the house itself.

With planning, a house can be oriented for natural ventilation. On a south-facing hill, breezes tend to move up the hill during day, and downhill at night. Near a large body of water, breezes blow from land to water at night.

3
WHEN YOU BUILD OR REMODEL

"In this whole industry of alternate energy, more people are being ripped off through ignorance than through malice."

William Enter
Inventor, Owner
Anabil Enterprises, Inc.

People who study such things say that half the houses which will be standing in the year 2000 have not yet been built. Between now and the end of the 20th Century, we will construct something like 40 million new housing units.

The accuracy of that projection depends a great deal on the general economic health of the nation's economy. The housing industry is among the more sensitive barometers indicating the U.S. economic and fiscal well-being. Inflation has boosted the cost of building materials, labor, and finished houses, until conventional single-family offerings are out of reach for many young couples, people approaching retirement age and a lot of us in be-

tween. Mortgage lending is among the first credit sources to dry up in periods of tight money and high interest policies.

But, unless we build better houses in the next 20 years than we have in the past 20, make them more energy efficient, they will continue to be drains on the dwindling energy reserves of the country and the pocketbooks of their owners. Even today, in many parts of the country, utility costs are higher than monthly mortage payments.

More and more prospective homeowners are taking an active role in planning and building their new homes, even to the point of doing a major part of the construction work.

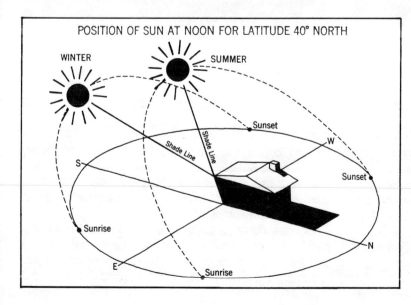

The seasonal elevation of the sun plus window exposures can influence your energy bills. The sun reaches windows placed on the south during the winter; it streams in windows facing east and west during the summer (Dept. of Agriculture).

Main living areas with sufficient south-facing glass area capture a lot of heat from solar radiation.

When the owner-builder does his homework well, the result can be more value and owner satisfaction per dollar spent. When *you* choose the site, the style of house, the utilities that will heat and power it, you not only have cut out several middlemen and their profit margins, you also have a better chance of owning a well made house that suits your family than if you buy a typical "built-to-sell" dwelling.

It starts with the location. Choose your location carefully; not just the site itself, but the area. Within the bounds of incorporated municipalities, backward building codes often stand in the way of inexpensive or energy-saving construction features. An attractive piece of rural land can carry hidden costs—water, roads, electrical service—that quickly erode any saving in the initial price of the site.

Any energy-saving design must start with the climate and building site in mind. We used to build homes with the climate as a major design factor. Even in prehistoric times, cliff dwellers located their homes to benefit from the full force of the winter sun, but where they were shaded in summer by overhangs. Climate was an important design factor in the square, stolid New England "salt boxes" and in the small, thick walled log cabins of the Northwoods.

The days of installing "brute force" petroleum heating systems to overcome the effects of poor design and sloppy building are over. It's high time we went back to designing houses with the climate and natural advantages of the building site in mind. An energy-conscious building program takes into account the location, configuration, orientation as to sun and prevailing breezes, layout or design of the living space, methods of construction and particular design details.

In chapters ahead, we point out how to plan and design the pieces—solar, wood, wind, water, etc.—that can be incorporated into a modern home. But the first order of business is to describe the basic package—the home and its immediate surroundings—that will let these alternate energy systems function to their fullest potential.

DESIGNING THE PACKAGE

People have different tastes in housing styles. Some like split-level ranch, some two-story colonial, some more contemporary designs. Despite style, however, the overall shape and volume of the home you build or remodel are major considerations in energy efficiency.

We all have to accept some compromises in home building, because of site limitations, financial limitations, or whatever. But you can gain almost complete control over your new home's environment to enjoy maximum comfort and still economize on both construction costs and energy consumption.

You may have trouble explaining some of the energy saving features you plan to incorporate in your home to loan officers, code officials and building inspectors. If you need help to get code and zoning approval of some of the cost saving ideas which we outline later in this chapter, contact a building contractor who has a good record at building low cost, energy efficient homes. Or, you may want to get in touch with the National Forest Products Association or the Western Plywood Association for examples of where these techniques are being used in the housing industry.

Here are some things to keep in mind:

Hot air rises. Remember this, if there are considerable temperature differences in a house, or in an individual room of the house. Also, heat flows from warm to cold surfaces or areas. Where walls and ceilings are well-insulated, but floors are not, the heat will be attracted and dissipated by the coldest surface.

Also, most people will feel chilled if their feet are cold, even if the average temperature at thermostat height is well into the comfort range. So, a minimum of six inches of insulation over crawl spaces and at least 1½ inches of polyurethane rigid foam around the perimeter of slabs can improve comfort and cut heating bills.

Incidentally, a concrete slab or a footing of pressure treated lumber can support a house as well as traditional basement or concrete footing-and-foundation structures. Depending on your home's design, these construction features may save money.

The earth itself provides some thermal resistance to heat flow, although the earth is not a particularly good insulator. If your building site slopes to the south or east, you may want to consider building the house below grade—either totally underground or with most of the wall area in contact with the earth. The major benefit of earth-contact houses comes from the earth's ability to moderate temperature changes. Earth slows down the temperature variations between interior and exterior, and provides protection from cold winter winds. In this way, the earth can contribute a great deal to reducing a building's heat loss. We'll have more on earth-contact shelters in Chapter 7.

One important thing to keep in mind is that with earth in contact with the house, you must use insulation on the outside of the structure—between the earth cover and the walls or roof of the building. This lets walls, roofs, ceilings and floors store heat that would otherwise be transferred to the surrounding soil.

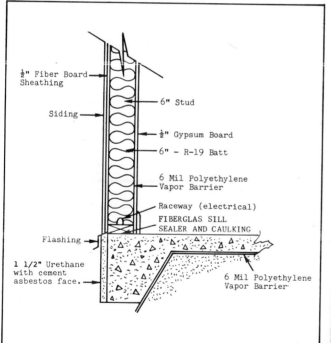

In some cases, the conventional concrete footing and foundation wall can be improved upon by using monolithic concrete slab or pressure-treated wood footings on tamped gravel beds.

Earth bermed against exterior walls to the level of window sills can save 10 to 15 percent; use vapor barriers to protect walls.

In your design, you may want to make use of a modified "earth-contact" principle and place bedrooms and bathrooms in the below-grade portion of the house (what is normally the basement). The amount of excavation need not be increased from that needed by a normal basement, but the first floor height could be raised to allow an 8 foot or 7′ 6″ ceiling height in the below-grade level, to accommodate two-foot-high clerestory windows in the "underground" floor level. This makes the space more livable than a typical basement and saves on both construction costs and energy.

Another approach to using earth is to berm it up against the walls of the house (after protecting walls from the moisture in the earth with a good vapor barrier). When conventional windows are used, say with three foot high sills, the earth can be banked to the first floor window sills. These approaches to underground or earth-contact shelters also reduce heat gain and require less cooling energy. If you plan to go to a total underground design, check the features in Chapter 7.

Window type, placement and area can be important to energy savings. Fifteen to 35 percent of the total heat loss can be attributed to windows.

It's good strategy to keep the window area to 10 percent or less of the total square footage in the living area. For example, a 1,600 square foot home should have no more than 160 square feet of windows.

By placing windows carefully, you can cut total window area to less than 10 percent. This lets you reduce heat loss and maximize solar gain without giving up much in natural lighting and outdoor "visibility." Put most of the windows on south-facing walls, to capture as much solar radiation as possible (regardless of the home heating system you install). With a good heat storage mass and insulating curtains or shutters, large south-facing win-

Casement windows allow less air infiltration than good quality double hung windows of the same size.

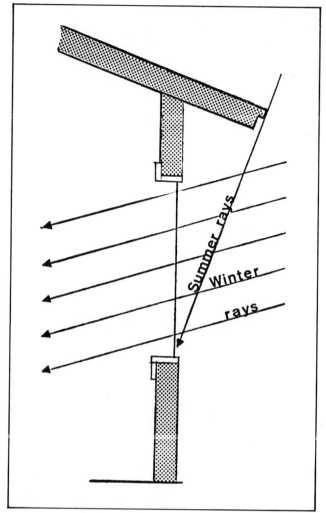

Roof overhangs can be built to shade windows when the sun is higher in summer, and still admit warming solar rays from winter's lower sun angle.

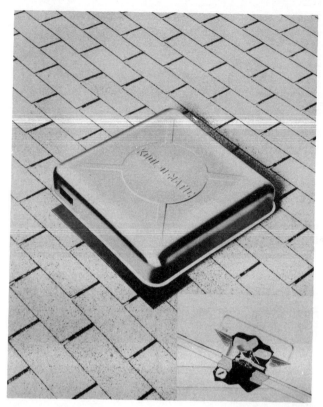

Powered roof vents can exhaust hot air from the attic. Install 80 square inches of soffit vent for each 100 CFM (cubic feet per minute) of vent fan capacity. (Courtesy: Kool-O-Matic)

dows can provide up to 50 percent of a building's space heating needs, in warmer climates.

The type of window you use can be important, too. Casement windows with good seals have about one-tenth the air infiltration of good quality double hung windows. Compared with a single pane of glass, dual pane windows with an air space between panes cut heat loss by 50 percent; adding another pane and air space, as with storm sashes or going to a triple glazed window, cuts heat loss by two-thirds.

In summer, windows can let heat into the house. To prevent excess heat gain in summer, windows should be shaded. Stopping the sun's rays before they hit the window is several times more effective than screening with drapes or curtains on the inside of the window. It's a fairly simple matter to design a roof overhang that will shade

the total glass area at noon during the hotter months of the year, yet allow the rays of lower winter sun to enter the window. This means you'll need to pay attention to the orientation of your house in relation to the sun's path through the sky at different times of the year.

Window frames built to accommodate insulating shutters are another good idea. And, if it's done during the initial construction, the shutter frames can be incorporated without having an "add-on" or afterthought appearance. Shutters not only screen out the sun during hot weather, they also can be closed over the glass at night and on cloudy days to prevent heat loss.

Power roof ventilators with eave vents evenly spaced along the soffit are more effective than gable louvers in ridding the attic of hot air buildup. Allow 80 square inches of soffit vent area for each 100 cubic feet per minute (CFM) of fan capacity. Power attic ventilators should have enough capacity to make 10 air changes per hour when running.

Speaking of the attic, don't locate any part of the heating or cooling equipment in the uninsulated part of the attic. An exception to this might be the collector of a solar system, where the collector itself is insulated. Another

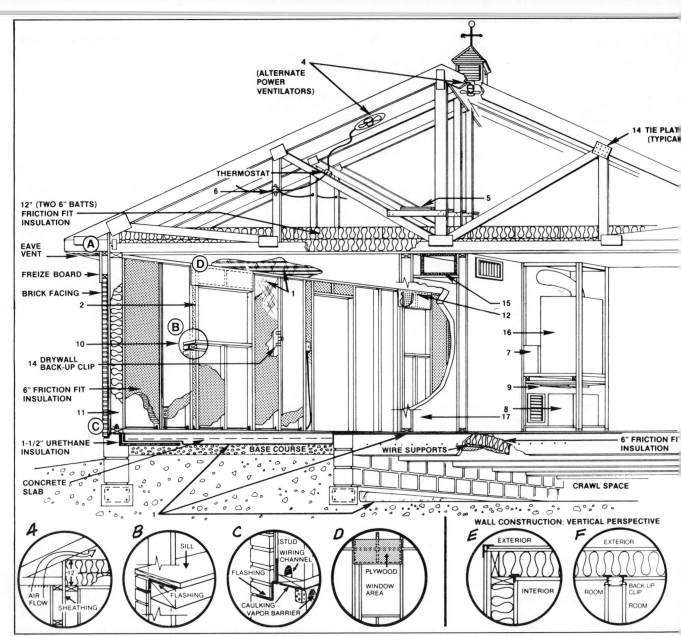

(ALTERNATE POWER VENTILATORS)

14 TIE PLAT(TYPICA

THERMOSTAT

12" (TWO 6" BATTS) FRICTION FIT INSULATION

EAVE VENT

(A)

FREIZE BOARD

BRICK FACING

2

(D)

1

15
12

16

(B)

10

7

14 DRYWALL BACK-UP CLIP

6" FRICTION FIT INSULATION

11

9

(C)

8
17

1-1/2" URETHANE INSULATION

BASE COURSE

WIRE SUPPORTS

6" FRICTION FI INSULATION

CONCRETE SLAB

CRAWL SPACE

1

5

6

WALL CONSTRUCTION: VERTICAL PERSPECTIVE

A
12
AIR FLOW
SHEATHING

B
SILL
FLASHING

C
STUD
WIRING CHANNEL
FLASHING
CAULKING VAPOR BARRIER

D
PLYWOOD
WINDOW AREA

E
EXTERIOR
INTERIOR

F
EXTERIOR
ROOM
BACK UP CLIP
ROOM

Design features of the Arkansas Energy-Saving House include: (A) ceiling insulation extends over stud wall to sheathing; (B) Window flashing drains through brick mortar joint; (C) base flashing extends from behind wall sheathing into course of bricks; (D) ½-inch plywood headers are glued and nailed in place of sheathing over windows; (E) arrangement of corner studs allow insulation to fill corner; and (F) non-bearing partitions join exterior walls with back-up clips.

exception is the air intake vent of a downdraft type oil or gas furnace.

In fact, if you build a tight, well insulated house and use any flame type space heating or water heating equipment, it's safer to locate the firing unit where it can be ducted with outside air through a six inch or larger round duct. Otherwise, fumes from the burner can be sucked into the living area at times, or the burner unit can partially deplete the living space of oxygen. Natural gas burners require 10 cubic feet of air for each cubic foot of gas burned.

In short, your goal in designing your home should be to gain as much control as possible over the environment

inside the home, at the lowest dollar cost and without sacrificing any comfort or safety.

The operating costs of a "conventional" home over the life of the mortgage can be far larger than the cost of insulation, a well designed natural heating system and other steps to reduce heating, cooling and power costs.

Once built, a home cannot economically be completely renovated for maximum energy efficiency.

COST-CUTTING CONSTRUCTION

In the construction of your new home, it's possible to build a house that is not only more energy efficient, but to build it faster and at a lower cost than conventional construction. As you design the new home, or major remodeling project, look into proven commercial construction techniques that are not widely used in home building: post-and-beam construction and the use of prefabricated trusses, to name a couple.

To start with, build no more house than you will need. Savings in material costs and heating bills can be considerably if you build only as much house as you need, and use all of the enclosed space efficiently.

Should you do your own prime contracting? If you have the skill and time for it, you may be able to get more for your money. A contractor charges 10 to 25 percent of the gross building cost for his overhead and profit—you can save a lot of this cost by subcontracting the trade work yourself. You'll need patience, and you'll become better at dealing with people if you do.

Be careful, though. Construction is full of rip-off artists who, either through ignorance or outright dishonesty, can waste a lot of your building investment. You may have to look around to find subcontractors willing to adjust their "normal way of doing it" to suit your cost cutting and energy saving requirements.

Here are some things to consider as you design and build your home, or remodel an existing one: Modified post-and-beam framing can save materials, compared with building traditional stud walls. And, if the post-and-beam structure is well designed and carefully built, the result can be both structurally strong and visually attractive.

Walls built of 2-by-6 studs on 24-inch centers cost less and go up faster than conventional 2-by-4 studs on 16-inch centers. The deeper wall cavity holds more insulation. (Owens-Corning Fiberglas)

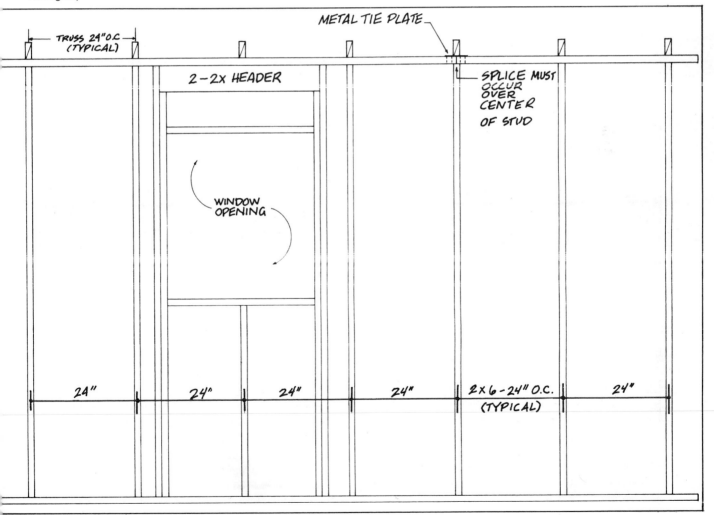

Even with a stud type building plan, you aren't necessarily married to 2x4 studs on 16 inch centers. A 2x6 stud construction 24 inches on center permits the use of R-19 insulation in exterior wall cavities, and saves construction time and money. In much of the country, 2x6 lumber in the shorter lengths used for studs is less expensive per board foot than 2x4 studs. Also, because of the larger cross-section of wood in a 2x6, you may be able to use No. 3 or Utility grade of most lumber species.

And you can save materials. Single plates on top of 2x6 stud walls can be used, eliminating the doubled "T" plate needed in conventional construction. Of course, you'll need to be sure that top plate joints, as well as rafters or roof trusses, rest directly over stud center lines and are well anchored to the wall stud below.

For roof framing, a modified truss, placed 24 inches on centers can save still more time and material. The trusses can be fabricated elsewhere and hauled to the site assembled. Often, pre-fabricated trusses cost less than the materials and time required to build them on site.

The U.S. Forest Products Laboratory has developed a pre-fabricated truss that uses only 2x4's. This framing system reduces the lumber needed to frame a conventional house roof by about 30 percent. The trusses can be placed on 24 inch centers rather than the conventional 16 inches.

The trusses for a 32 foot span weigh only 250 pounds each. When tied to wall studs with an anchor at each end, this method of construction makes a rigid structural unit that better stands wind, snow loads and other stresses than conventional framing.

If you're designing a house with a cathedral ceiling, you can still use trussed roof framing, but will need to go to a scissors type truss with extended legs. If you buy trusses from a reputable fabricator, he will be able to construct the unit that best suits your design and conditions. If you will be making up your own trusses, either get an experienced framer to build a pattern or study a good reference on truss design before you start cutting and nailing.

With truss roof framing, no partition walls need to bear part of the roof load. This gives more flexibility in laying out the interior of the house. Also, since partitions have to support only their own weight, you can frame them of 2x3 lumber on 24 inch centers—another savings in time and material.

You can reduce framing costs still more by using back-up clips to install sheet rock or paneling, and eliminate the need for T framing where walls join. Panel back-up clips also leave more room for insulation in exterior walls.

The idea of using prefabricated trusses works equally well in floor framing. In fact, well designed bridge type floor trusses can eliminate the need for under-floor beams and lally columns in some construction. Trusses also make the installation of wiring, plumbing and heating ducts easier and faster than with conventional joist framing.

With 2x6 studs, you can pre-cut raceways to hold the electrical conduit and plumbing pipes out of the way of the insulation, through the bottom of each stud before it is installed. This also saves the electrical and plumbing contractors several hours of hole cutting time.

To save more on materials, you may want to ask the electrician about the advisability of using fewer electrical circuits but running heavier conductor wire. To avoid compressing the insulation in exterior walls any more than necessary, you may want to hold electrical switches and outlets to a minimum on these walls—perhaps, even run the circuits in exposed raceways on the inside of the wall covering. Caulk or tape all electrical outlets to

High ceilings and cathedral ceilings make for cooler rooms in summer, but are heating liabilities in winter. A movable insulated ceiling panel can reduce ceiling height to eight feet during the heating season. (National Solar Heating and Cooling Center)

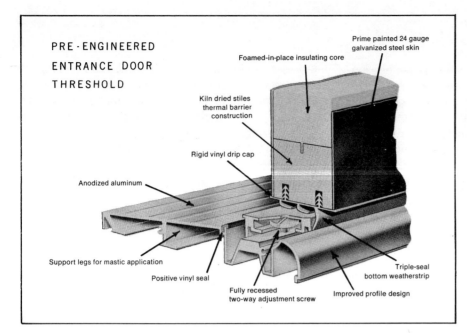

PRE·ENGINEERED
ENTRANCE DOOR
THRESHOLD

Prime painted 24 gauge
galvanized steel skin

Foamed-in-place insulating core

Kiln dried stiles
thermal barrier
construction

Rigid vinyl drip cap

Anodized aluminum

Support legs for mastic application

Positive vinyl seal

Fully recessed
two-way adjustment screw

Improved profile design

Triple-seal
bottom weatherstrip

Prehung thermal doors feature a polyurethane foam core sheathed in 24-gauge metal, with magnetic refrigerator-type seals and adjustable bottom weatherstripping (Stanley Door Systems)

the sheetrock or other wall covering to cut down on air infiltration.

Use ventless hoods with charcoal filters over the kitchen range, rather than cut a hole through the exterior wall. Also, don't place such fixtures as bathroom medicine cabinets in exterior walls where they would interfere with insulation.

By confining the plumbing to a relatively small rectangle to serve kitchen and bathrooms located in the same vicinity, you not only reduce the amount of piping materials and time needed to install plumbing, but also reduce the length of hot water lines and cut heat losses.

If you're going to use a fireplace, put in an efficient one, such as a Heatilator or Preway Energy Mizer. These units have glass front doors and use combustion air ducted from outside the house. They also are provided with dampers to stop air infiltration when the fireplace is not in use.

Generally, it's better to use unfaced insulation in walls and ceilings, then install a separate vapor barrier. Use 6 mil polyethylene over the entire ceiling and wall interior. Or, as an alternative, use foil-backed gypsum board. This is more effective than using insulation faced with foil or Kraft paper; the inside surfaces of studs and ceiling joists are also covered with the moisture barrier. Use 6 mil plastic as a ground cover in crawl spaces, also.

It may seem unusual to find several pages on cost cutting construction techniques in a book that deals mostly with natural energy systems. But it fits, if you stop to think about it. Lower construction costs, because of less material used, fewer pieces to fit and install and shorter building time, can help pay part of the initial cost of a natural heating or power system. These techniques also save money in maintenance and repair costs over the life of the house.

PLANNING THE ENERGY SYSTEM

In this specialized age, we like to look for a single solution to a problem; one cure-all that will end our woes. But no one source of natural energy will replace the petroleum we have come to depend upon so heavily during the past several years.

And, it's unlikely that any one source of natural energy can supply all the heat and power you will need to make your home liveable and comfortable. For one thing, the supply of most natural energy resources is not continuous: the wind blows intermittently; the sun doesn't shine every day.

For another thing, natural energy comes in different forms: solar as light; wind and water as mechanical; wood as stored heat. These sources of energy can most efficiently be used in the form they normally take, or with only one or two conversion steps. Solar energy can be converted to heat fairly easily; the mechanical energy of the wind or water can be converted to electrical energy. The more complex the conversion process, the more expensive the equipment to do it.

You have heard arguments, probably, that such natural systems as solar heating and wind generated electricity are not yet practical for the homeowner to buy or build. And, if we could be sure that we'd have an unlimited supply of petroleum at even today's prices, that might be true.

But we have no such assurances. In fact, the world's oil situation is increasingly glum. And, while it would be cost prohibitive for most of us to build systems to supply *all* of our heat and power, it is well within the reach of most of us to install systems that will produce a good percentage of the energy we need in our homes.

New designs of wood burning stoves and furnaces bring the efficiency of these natural energy users to within a few percentage points of oil and gas burners. Solar sys-

45

tems that provide half or more of a home's winter heating needs now are proven, workable energy sources. Wind generating plants are more efficient at producing electricity out of low speed winds; batteries are more efficient at storing current for those times the wind does not blow.

So, the argument that alternate, or natural, energy sys-

tems are not practical is based on some faulty generalizations. In all probability, the hardware needed to collect, store and use natural energy will become less expensive as the techniques of mass production and mass marketing are brought to bear on those industries producing the equipment. Look what happened with electronic pocket

"Degree Days" per Month (for selected locations)

Location	July	Aug.	Sept.	Oct.	Nov.	Dec.	Jan.	Feb.	Mar.	Apr.	May	June	Year
Birmingham, AL	0	0	6	93	363	1555	1592	1462	1363	108	29	20	12,551
Mobile, AL	0	0	0	22	213	1357	1415	1300	1211	142	20	20	11,560
Anchorage, AK	245	291	516	930	1284	1572	1613	1316	1293	879	592	315	10,864
Flagstaff, AZ	46	68	201	558	867	1073	1169	1991	1899	651	437	180	17,152
Phoenix, AZ	0	0	0	122	234	1415	1474	1328	1217	175	20	20	11,765
Little Rock, AR	0	0	9	127	465	1716	1756	1577	1434	126	28	20	13,218
Eureka, CA	270	257	258	329	1414	1499	1546	1470	1505	438	372	285	14,643
Oakland, CA	53	50	45	127	1309	1481	1527	1400	1353	255	180	190	12,870
San Djego, CA	6	0	15	137	1123	1251	1313	1249	1202	123	184	136	11,439
Denver, CO	6	9	117	428	1819	1035	1132	1938	1887	558	288	166	16,283
Pueblo, CO	0	0	54	326	1750	1986	1085	1871	1772	429	174	115	15,462
New Haven, CT	0	12	87	347	1648	1011	1097	1991	1871	543	245	145	15,897
Wilmington, DE	0	0	51	270	1588	1927	1980	1874	1735	387	112	26	14,930
Jacksonville, FL	0	0	0	112	1144	1310	1332	1246	1174	121	20	20	11,230
Miami Beach, FL	0	0	0	20	30	240	256	236	39	20	20	20	141
Tallahassee, FL	0	0	0	128	198	1360	1375	1286	1202	136	20	20	11,485
Atlanta, GA	0	0	18	127	1414	1626	1639	1529	1437	168	125	20	12,983
Savannah, GA	0	0	0	147	1246	1437	1437	1353	1254	145	20	20	11,819
Boise, ID	0	0	132	415	1792	1017	1113	1854	1722	438	245	181	15,809
Chicago, IL	0	0	81	326	1753	1113	1209	1044	1890	480	211	148	16,155
Springfield, IL	0	0	72	291	1696	1023	1135	1935	1769	354	136	118	15,429
Fort Wayne, IN	0	9	105	378	1783	1135	1178	1028	1890	471	189	139	16,205
Evansville, IN	0	0	66	220	1606	1896	1955	1767	1620	237	168	20	14,435
Des Moines, IA	20	29	199	363	1837	1231	1398	1163	1967	489	211	139	16,808
Topeka, KS	20	20	157	270	1672	1980	1122	1893	1722	330	124	112	15,182
Wichita, KS	20	20	133	229	1618	1905	1023	1804	1645	270	187	26	14,620
Louisville, KY	20	20	154	248	1609	1890	1930	1818	1682	315	105	29	14,660
New Orleans, LA	20	20	20	119	1192	1322	1363	1258	1192	129	20	20	11,385
Shreveport, LA	20	20	20	147	1297	1477	1552	1426	1304	181	20	20	12,185
Caribou, ME	178	115	336	682	1044	1535	1690	1470	1308	858	468	183	19,767
Portland, ME	112	153	195	508	1807	1215	1339	1182	1042	675	372	111	17,511
Baltimore, MD	20	20	148	264	1585	1905	1936	1820	1676	327	190	20	14,651
Boston, MA	20	29	160	316	1603	1983	1088	1972	1846	513	208	136	15,634
Detroit, MI	20	20	187	360	1738	1088	1181	1058	1936	522	220	142	16,232
Sault Ste. Marie, MI	196	105	279	580	1951	1367	1525	1380	1277	810	477	201	19,048
Int. Falls, MN	171	112	363	701	1236	1724	1919	1621	1414	828	443	174	10,606
Minneapolis, MN	122	131	189	505	1014	1454	1631	1380	1166	621	288	181	18,382
Jackson, MS	20	20	20	165	1315	1502	1546	1414	1310	187	20	20	12,239
St. Louis, MO	20	20	160	251	1627	1936	1026	1848	1704	312	121	115	14900
Billings, MT	26	115	186	487	1897	1135	1296	1100	1970	570	285	102	17,049
Glasgow, MT	131	147	270	608	1104	1466	1711	1439	1187	648	335	150	18,996

calculators and digital watches in just a few years. But you must weigh that probability against the energy costs you will incur with conventional heating and power systems this year and next.

The starting point for planning the operating plant for your own home is to compute your family's energy re-

quirements for heat, electrical power, mechanical power, etc.

Let's consider space heating first, since home heating is the big energy requirement for most of us. Determining the heating needs of a house is a complicated undertaking, and depends upon several factors. But, basically, you

"Degree Days" per Month (for selected locations)

Location	July	Aug.	Sept.	Oct.	Nov.	Dec.	Jan.	Feb.	Mar.	Apr.	May	June	Year
Lincoln, NE	20	26	175	301	1726	1066	1237	1016	1834	402	171	130	15,864
Valentine, NE	29	112	165	493	1942	1237	1395	1176	1045	579	288	184	17,425
Ely, NV	128	143	234	592	1939	1184	1308	1075	1977	672	456	225	17,733
Las Vegas, NV	20	20	20	178	1387	1617	1688	1487	1335	111	26	20	12,709
Concord, NH	26	150	177	505	1822	1240	1358	1184	1032	636	298	175	17,383
Atlantic City, NJ	20	20	139	251	1549	1880	1936	1848	1741	420	133	115	14,812
Albuquerque, NM	20	20	112	229	1642	1868	1930	1703	1595	288	181	20	14,348
Raton, NM	29	128	126	431	1825	1048	1116	1904	1834	543	301	163	16,228
Buffalo, NY	119	137	141	440	1777	1156	1256	1145	1039	645	313	199	17,062
New York, NY	20	20	127	223	1528	1887	1973	1879	1750	414	124	26	14,811
Raleigh, NC	20	20	121	164	1450	1716	1725	1616	1487	180	134	20	13,393
Bismarck, ND	134	128	222	577	1083	1463	1708	1442	1203	645	329	117	18,851
Akron, OH	20	29	196	381	1726	1070	1138	1016	1871	489	202	139	16,037
Cincinnati, OH	20	20	154	248	1612	1921	1970	1837	1701	336	118	29	14,806
Columbus, OH	20	26	184	347	1714	1039	1088	1949	1809	426	171	127	15,660
Oklahoma City, OK	20	20	115	164	1498	1766	1868	1664	1527	189	134	20	13,725
Portland, OR	125	128	114	335	1597	1735	1825	1644	1586	396	245	105	14,635
Philadelphia, PA	20	20	160	291	1621	1964	1014	1890	1744	390	115	112	15,101
Pittsburgh, PA	20	29	105	375	1726	1063	1119	1002	1874	480	195	139	15,987
Providence, RI	20	116	196	372	1660	1023	1110	1988	1868	534	236	151	15,954
Columbia, SC	20	20	20	184	1345	1577	1570	1470	1357	181	20	20	12,484
Sioux Falls, SD	119	125	168	462	1972	1361	1544	1285	1082	573	270	178	17,839
Nashville, TN	20	20	130	158	1495	1732	1778	1644	1512	189	140	20	13,578
Amarillo, TX	20	20	118	205	1570	1797	1877	1664	1546	252	156	20	13,985
Dallas, TX	20	20	20	162	1321	1524	1601	1440	1319	190	26	20	12,363
Houston, TX	20	20	20	26	1183	1307	1384	1288	1192	136	20	20	11,396
Salt Lake City, UT	20	20	181	419	1849	1082	1172	1910	1763	459	233	184	16,052
Burlington, VT	128	165	208	539	1891	1349	1513	1333	1187	714	353	190	18,268
Richmond, VA	20	20	136	214	1495	1784	1815	1703	1546	219	153	20	13,865
Seattle, WA	150	147	129	329	1543	1659	1738	1599	1577	396	242	199	14,426
Spokane, WA	29	125	168	493	1879	1083	1231	1980	1834	531	288	135	16,656
Charleston, WV	20	20	163	254	1591	1865	1880	1770	1648	300	196	29	14,476
Green Bay, WI	128	150	174	484	1923	1333	1494	1313	1141	654	335	199	18,028
Milwaukee, WI	143	147	174	471	1876	1252	1376	1193	1054	642	372	135	17,635
Cheyenne, WY	119	131	210	543	1924	1101	1228	1056	1011	672	381	102	17,278

The "heating degree day" method of estimating heating loads has been in general use by the heating industry for about 30 years, and is still the simplest method of figuring how much heat will be needed. The base temperature, as mentioned above, is 65 degrees F. Normally, heating is not required when the outside temperature is 65 or above.

Heating degree days are cumulative. And, where heating by petroleum fuels are concerned, there is a linear relationship between degree days and fuel consumption. For example, if you live in Milwaukee, you can expect to burn nearly three times as much heating fuel in December as in October.

need to know how much of the year you will need to use some heat.

One way to compute this is in "degree days." This is determined by subtracting the average outside temperature over a 24 hour period from a base of 65 degrees F. For example, if the average temperature for October 30 is 45°, the "degree days" for that particular day would be 20. To find the degree days for a month or year, you simply add together all the degree days for that period.

See the chart on pages 46 and 47 for the degree days per month for some selected locations around the U.S.

You may be able to install a natural home heating system that provides all the heat you'll need. If you have an ample firewood supply and will have someone at home to stoke the stove now and then through the heating season, wood heat can supply your space heating needs adequately.

But it's more likely that the natural energy system (or systems) you employ will need some sort of "back-up" conventional heat. This may be particularly true of solar installations in many areas, where another source of heat is needed for cold, cloudy periods.

When you're shopping for a supplemental heating system, keep in mind these considerations:

1. The "back-up" system should be compatible with the natural heating plant, as noted earlier.

2. The heating system should use a fuel or heat source that is expected to be available for some time in the future.

3. If several conventional heating fuels are equally available, choose the one with the lowest cost, but base your estimate of fuel costs on the *actual heat* from the fuel, rather than on the available BTUs in the fuel.

For example, the available heat per unit in BTUs (British Thermal Units—one BTU is the amount of heat energy required to raise the temperature of one pound of water one degree F. at sea level) may be quite different from the actual heat recovered from that fuel (see chart).

In most areas of the country, more than one heating

An example of modern wood-burning equipment is this deluxe imperial circulator which not only can heat two rooms or one large room, but can be placed only 12 inches from the wall. (Courtesy: Ashley Wood Heaters)

fuel will be available; however, prices vary widely. You can investigate prices of various fuels in your area and use the chart below to compute the actual heating value of each.

For example, based on the chart, which is the better buy—heating oil at 80 cents per U.S. gallon, or natural gas at 55 cents per Therm (100 cubic feet)? The natural gas is the better heating buy, in this case. A penny spent for natural gas will buy 1,273 BTUs of actual heat; while a penny spent for fuel oil only buys 1,225 BTUs.

There are some "side effects" to certain heating fuels which may be important to some people, although they are hard to put a dollar value on. For instance, heating with coal or wood is usually "messier" and takes more time and effort than using gas or electricity. However, if you can cut and haul 11 million BTU's of heat in a cord of

Fuel Type	Unit	BTU's Available	System Efficiency	Actual BTU's of heat
Heating Oil	U.S. Gal.	140,000	70%	98,000
Propane (L.P. gas)	U.S. Gal.	100,000	70%	70,000
Natural Gas	Therm	100,000	70%	70,000
Coal, Anthracite	Ton	25,400,000	60%	15,240,000
Coal, Bituminous	Ton	26,200,000	60%	15,720,000
Electricity	KWH	3,412	100%	3,412
Wood, Oak	Cord	22,000,000	50%	11,000,000

PROJECTING ANNUAL POWER USE

Projecting your annual use of electrical power can be even trickier than figuring your home heating requirements. The chart below lists the annual energy requirements of household electric appliances and equipment, based on average use. To use these figures for your own projections, you'll need to take into account your family's use habits and the geographical region of the country.

Appliance	Wattage	Estimated annual kilowatt/hours	Operating Cost (@ .05/KWh)
Central air-conditioner (Based on 500 hours of operation per year)	5,000-8,000	3,250	$162.50
Room air-conditioner (Based on 1,000 hours of operation per year)	860	860	$ 43.00
Clothes dryer	4,000-8,000	990	$ 49.50
Dishwasher	700-1,000	365	$ 18.25
Freezer (16 cu.ft.)	300-500	1,200	$ 60.00
Range-oven	8,000-15,000	700	$ 35.00
Refrigerator (17.5 cu.ft.) (Frost-free type)	200-300	2,250	$112.50
Automatic clothes washer	400-600	110	$ 5.50
Water heater	2,000-5,000	5,200	$260.00
Electric blender	250	15	$ .75
Coffee maker	600-1,000	140	$ 7.00
Mixer	150-250	15	$ .75
Microwave oven	Various	190	$ 9.50
Toaster	200-500	40	$ 2.00
Trash compactor	300-500	50	$ 2.50
Electric blanket	150-300	150	$ 7.50
Dehumidifier	250-400	380	$ 19.00
Attic fan	700-950	300	$ 15.00
Window fan	100-200	170	$ 8.50
Iron (hand)	650-1,200	150	$ 7.50
Stereo record player	100-300	110	$ 5.50
Color television— Tube type	200-500	660	$ 33.00
Solid state	100-200	440	$ 22.00
Vacuum cleaner	200-400	45	$ 2.25

wood for $20 per cord, that's the equivalent of the actual heat in $97.60 worth of fuel oil at 80 cents per gallon. The $77.60 difference would pay for some of the inconvenience associated with burning wood.

Would a generating plant that provides half of your electrical needs be worthwhile?

You will need to evaluate any alternate/natural energy system on the basis of its benefit to your family and the house you build or remodel. You can do that, partially, by comparing natural energy systems with conventional systems on the same basis you compare buying and renting a house.

When you buy, you slowly but surely gain ownership. And you assume such costs of ownership as taxes, insurance, maintenance and repair. On the other hand, you gain some tax deductions and credits that are not provided renters, plus whatever appreciation in value the property earns.

As a renter, you do not have to worry as much about repairs, maintenance and property taxes, but when it comes time to move, all those rental dollars are gone forever.

Buying and installing natural energy systems is much the same kind of proposition. You incur initial costs and lifetime operating costs, but when the last payment is made, the system is there to provide low cost or free heat or power. If you rely on a conventional system altogether, it's like renting—there's no hope of ever regaining those dollars spent for fuel and electricity.

When you evaluate any natural energy system, you need to consider on the debit side initial costs, anticipated annual maintenance and operating costs and interest on your capital. On the plus side, you have annual energy savings over a conventional system, appreciation of the house and its energy systems, tax breaks from federal and state governments and tax deductions for the interest payments you make. Another big question is energy inflation: how much will the cost of conventional energy increase over the life of the system?

The pivotal answer in your calculations will be the amount of time you will need to recover your investment. If the payback period for a natural energy system is, say 10 years, and you have a sound, durable installation, you can be pretty sure the system will hold up for that long before it needs major repairs or replacement.

The big drawback to answering that question, however, is how much of the load can be picked up by natural energy sources? In each of the sections ahead, we outline how to compute the heat or power output from the different equipment and systems discussed. Still, estimating closely what percentage a particular solar installation will provide for your winter heating needs is at best fraught with chances of error.

The big obstacles to satisfaction with any new product or system are these:

1. The buyer's lack of knowledge and experience.
2. Manufacturers and installers who unintentionally build shoddy products and install them carelessly.
3. Deliberate fraud and misrepresentation.

The best way to side step all three of these obstacles is for the homeowner to recognize his own limitations, and to rely upon *competent* engineering counsel. But here again, much of the alternate energy business is still in its infancy—not all engineers are experienced in natural energy systems. You'll want to know that your "expert" actually knows his business.

You'll want to check out manufacturers, dealers, installers and other people you'll need for help with your building project. This book, and others on the subject, are at best basic guidelines; sources of ideas and how to apply them. Beyond these pages, you will still need some professional, on-the-spot counsel. Here's how to make sure you get your money's worth in good advice:

• Don't be afraid to ask for proof that a product or piece of equipment will perform up to claims. Ask for a report of testing by an independent laboratory or public university—the manufacturer isn't necessarily dishonest, but he is not in a position to be absolutely objective about his product.

Equipment warranties should spell out repair and replacement terms: Who pays for what, and for how long?

- Make sure the product or equipment is covered by a warranty. If it's a "limited" warranty, know what the limits are. Who provides the service or repairs, should the equipment need it?
- Talk to people who own the system or one like it. The manufacturer or seller should be able to provide a list of previous buyers.
- Check with your local builders' association, consumer service office or Better Business Bureau for any complaints against the manufacturer, dealer or installer of equipment. Be suspicious of businesses that use only a post office box number in their address. This doesn't mean the firm is dishonest—many legitimate businesses use postal boxes—but it's a common tactic of fly-by-night operators who may need to seek new territory in a hurry.

Your best bet is to buy as much material and equipment as you can from a reliable dealer, get a good warranty and nail the installer down with a contract that makes him come back to fix any mistakes he made.

Line up the best help you can find to design and lay out the heating system supply. With a forced draft heating system, flared duct inlets and "hard" bends in the sheet metal designed with a venturi-type angle can help air flow smoothly. The same holds true for hot water supply piping. Use one size larger pipe or copper tubing to overcome some resistance to flow in the line. These techniques can let you step down a size in the electric motor needed to operate the system. For example, you may be able to use a ⅓ horsepower motor rather than a ½ horse motor. This not only saves costs in building, but saves operating costs from now on.

Some preliminary tests show that air blower operation can cost from $12 to $70 per year in the typical house. So, a more efficient blower with a small motor will rapidly make a big change in energy bills.

INSIDE EQUIPMENT

When buying appliances for your new home, comparison shop. Compare initial cost, energy use information and operating costs of similar models. The dealer should be able to give you the wattage of appliances, and provide an average estimate of total power consumed in a year's time by appliances such as home freezers and refrigerators.

Before you buy a new appliance with special features, find out how much energy it uses compared with a more conventional model.

If you install central air-conditioning, look into models with "air-economizer" features. This system turns off the compressor and circulates outside air through the house when it's cooler outside than it is inside. By using cooler outside air, the machine reduces the drain on electricity. Check the energy efficiency rating (EER) of window and through-the-wall type room air conditioners. A unit with an EER of 4 will cost about three times as much to operate as one with an EER of 12, but may cost about the same to buy.

You'll save money if you install gas ovens or ranges with automatic (electronic) ignition, rather than pilot lights. This will save up to 41 percent of the gas used in a conventional oven and 53 percent of the gas used in top burners.

A refrigerator or deep freeze will cost as much or more to operate over its lifetime as the unit costs initially—it pays to buy one that is energy efficient. Most refrigerators and freezers have heating elements in walls and doors to prevent "sweating" on the outside of the appliance in hot, humid weather. However, these elements are needed only in humid weather, and some units have "power-saver" control switches to turn off the heater when it isn't needed.

Consider buying refrigerators and freezers that have to be defrosted manually. Although they take more effort to defrost, these appliances use considerably less energy than automatic defrost models. Incidentally, chest type freezers take up more floor space but lose less cold air than upright ones when the door is opened.

Pumps, blowers and other operating hardware should be designed for the job. For example, circulating pumps in solar or boiler hot-water systems should be built for hot-water service. (Courtesy of Grundfos Pumps Corp.)

Energy-saving appliances can save electrical power and dollars. Refrigerator-freezers with extra insulation and power-saver switches can save $22 to $43 dollars per year. (Courtesy of Sears, Roebuck and Company)

An "instant-on" television set uses energy all the time to keep its innards warmed up—that's why it comes on instantly. If you buy a TV with this feature, you may want to install the set where it can be plugged into a wall outlet controlled by a switch—or, install an on-off line switch between the wall outlet and the TV set's cord.

Plan your home lighting sensibly. Reduce lighting where possible, concentrating it in work areas or where reading is done. Fluorescent bulbs rather than incandescent lights should be used wherever possible.

Ordinarily, lighting doesn't consume a big percentage of the electricity used each month. However, in a year's time, it costs unnecessary dollars to provide more light than is needed.

If you are considering an experimental entry into wind or water generated electrical power, you may want to install a small 12 volt generating plant—perhaps an automobile generator, voltage regulator and battery—to power outdoor lights and other 12 volt equipment. The battery will store enough power to run the lights for much of the time the generating plant is not turning, and you can buy a low amp battery charger to operate the system during longer periods when you cannot generate power from natural sources. Or, you can install a "step-down" transformer and a 12 volt lighting system outdoors that makes very little drain on house current.

In general, it's a good idea to provide as much light as is needed, where it's needed. As we mentioned earlier, light intensity varies with the *square* of the distance from the source. When you're reading, one 75 watt bulb over your left shoulder puts more light on the page than a bank of lights hidden along the edge of the ceiling.

The wattage of a light bulb refers only to the amount of electricity the unit consumes—it is not much of an indicator of light intensity. The measure of light or brightness emitted by a bulb is *lumens* which may be quite a different thing from watts. For example, one 150 watt bulb provides 2,880 lumens (when new), while two 75 watt bulbs offer only 2,380 lumens—although both consume the same amount of electricity.

When planning the lighting in your home, do it by lumens, rather than by watts. You can save money during installation and in electricity comsumption. Here are recommended lumen levels for typical about-the-home tasks:

Visual task	Lumens
Reading and writing	70
Kitchen work, at sink	70
Kitchen work, at range	50
Laundry chores	50
Sewing—dark fabrics	200
Sewing—light fabrics	100
Playing cards, chess, etc.	30
Shaving, grooming	50
Most shop work	50
Relaxing, watching TV	10

LANDSCAPE FOR ENERGY

You can influence your energy bills with careful landscaping, too. When the house and its surrounding grounds are combined in the total energy saving package, the result can be an energy efficient, easy-on-the-eyes homestead.

The landscape plan you choose will depend on your climate, the topography of your residence and—to some extent—on the natural energy features you incorporate. For instance, if you erect a wind generating plant, you do not want to plant tall trees that might deflect the breeze over or around the windmill.

Sunlight hours, prevailing winds in different seasons, the angle and intensity of the sun at different times of the year—all of these elements vary from region to region, but all will influence your total landscaping plan.

You can mass deciduous trees in an arc that curves from southeast to southwest, to shade the house and lower summer cooling costs. These trees lose their

With the variety of lights available many different effects can be achieved economically. (A) With all lighting systems on full, this room is well lighted but without glare. Downlights over fireplace are low wattage incandescent lamps, while paneling is lighted with warm white fluorescents behind cornice board. Recessed in the ceiling are low wattage incandescent spotlights.

All can be dimmer controlled. Table lamp is on a three-way switch. (B) Brick wall is highlighted by ceiling spots while other lightsare dimmed. (C) With just a 75-watt spotlight over flowers on table, and two more directly over front of chairs, room is conversationally lighted yet visually enlarged. (General Electric Company)

leaves in autumn, and permit the lower winter sun to operate a solar system or help warm the house through south windows. However, if you use a solar water heating system year 'round, you may want to only partially screen the sun with trees and use some other shading method for windows in summer.

If your property has enough space, an old-fashioned windbreak of evergreen and deciduous trees and shrubs can help block winter winds. If space is limited, a high fence on the north and west sides of the house will deflect the wintry blast.

Ground covers along the south side of the house cut down on reflected light from summer sun; whereas concrete walkways and patios close to the house on south and west sides can reflect a lot of heat into the house.

The best time to plan the weather wise and energy smart features of your home is in the "pencil-and-paper" stage. It's hard to have too much information before you start building. Talk with homeowners who are using systems you are considering. You can learn a lot of good things—and things not to do—from those who have gone before in this area of natural energy.

Energy-saving plans should extend to the home grounds, where carefully chosen, well-located trees and shrubs can contribute to comfort and energy efficiency year-round.

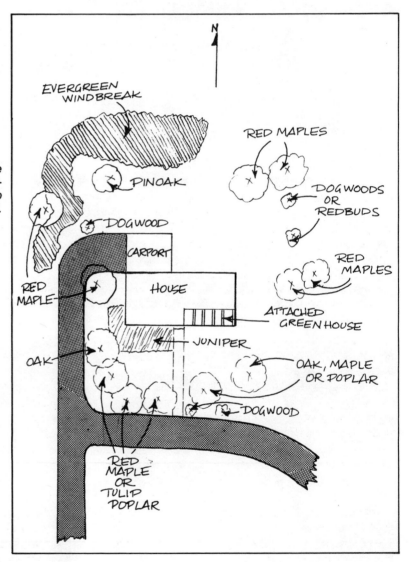

4
SOLAR ENERGY IS A COLLECTOR'S ITEM

"Americans are seeking ways of using the sun as an inexhaustible source of clean energy."

President Jimmy Carter
The White House
Washington, D.C.

Have you ever gotten into your car after leaving it parked in the sun with the windows up? If you have, you probably noticed how warm the interior of the automobile was, even if the day was raw and wintry.

That's solar energy at work.

While an automobile is not the most efficient method for collecting, storing and using heat from the sun, the principle is the same as with the most sophisticated solar heating set up. A transparent surface lets the sun's radiant energy through, the solar light turns to heat when it strikes objects and surfaces behind the glass, and the longer heat waves are trapped to warm the interior space.

Every day, the world receives 10,000 times more energy from the sun than mankind produces from all conventional fuels combined. The trouble is, although sunshine is free, its pipeline costs.

Solar energy for home heating and domestic water heating is here now, and the industry is among the faster growing businesses in the country. Whether it will help you, individually, in terms of producing real savings in energy costs, depends on a number of things. Where do you live? What kind of house do you have or plan to build? How much and what quality insulation does your home have? What kind of heating system do you presently use, and what does its fuel cost?

Solar space heating has been used in a home in Coral Gables, Florida, since 1938. The owner figures he has saved well over $10,000 in electric bills in that time, and the original collectors are still in place and working as well as when they were installed more than 40 years ago.

SHOULD YOU GO SOLAR?

Despite advertising claims by manufacturers of solar equipment, solar heating is *not* for everyone. Certainly not right now.

For your specific area, you will need to know how much usable radiation falls during the times you need to use a solar system. This figure, along with the system's estimated efficiency and the size of the collector area, will help you determine how many BTUs of heat you can expect to receive from any particular system—and whether the installation of the system is worth the cost, in terms of the BTUs per dollar invested.

Also, ask yourself whether your house or lot can handle a collector and storage system large enough for your needs. The collector will need to face a southerly direction and receive unobstructed sunlight.

What happens if your neighbor plants a tree that shades your entire collector? (This whole area of "solar rights" is open range for litigation in most states. One state has passed a law preventing the shading of a neighbor's solar collector; other states no doubt will enact solar legislation before long.)

Where you live is of key importance in deciding what kind of solar installation to build—or whether to invest in solar heating at all. In winter, the people who live in Maine do not receive as much sun as people who live in Florida. A 1,500 square foot house in Bangor would need considerably larger equipment to collect and store solar heat to accomplish the same percentage of total heating as one in Miami. However, the economics of solar heating are better in Maine than in Florida, because the solar energy that is collected can be used to offset a much larger winter heating bill.

There are four main uses of residential solar heating: heating the home, heating water for domestic use, heating *and* cooling the home and heating a swimming pool. There are combinations of these end uses. A system could be used to heat the house in winter, to heat domestic hot water year round and to heat a swimming pool in late spring and early fall.

For cost effectiveness, most solar systems are not designed to handle all the anticipated demand. If a solar system were designed to handle 100 percent of the heating load during the coldest part of winter, for example, the system would have to be sized to meet the demand for limited periods—the extra cost normally is not justified by the savings in conventional fuel.

Understanding how a solar system works is the beginning of wisdom in your decision as to whether a unit can be practical for your home.

Ask yourself these questions:

1. Is it suitable for my house and location?
2. Is it a worthwhile investment that will repay my first costs in a reasonable time?
3. Can I afford it right now?

What is a smart solar purchase for a homeowner in one area of the country may be foolish for another in another area. You don't have to be an architect or mechanical engineer to make a good buy in solar equipment. In fact, not every architect or engineer around the country has a thorough understanding of solar systems. Unfortunately, not all of the "experts" who hang out their shingles as solar consultants are as qualified as they could be.

At some point in your planning, you probably will want to call on expert advice, and we'll talk more about how to find and evaluate contractors later on. But before you get to that stage, you can do a lot to decide on what type—if any—solar heating system best suits your needs.

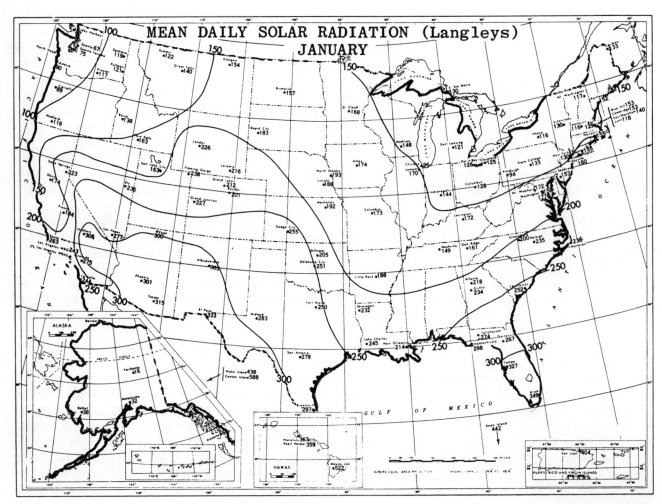

Courtesy of U.S. Dept. of Commerce

Site orientation is important, but not vital. If collectors are to be mounted on the building, the long sides of the house should face north and south. In most parts of the country, the prevailing winter winds are from the west or northwest; so a narrower west wall cuts infiltration from chilly breezes.

Even more important is the effect of the summer sun. The west wall picks up the greatest heat gain in summer, so the less area broadside to the lowering west sun has benefits in both winter and summer.

During the winter, the sun travels low in the southern sky from east to west. Windows facing south tend to act as solar collectors to pick up a great deal of the sun's heat.

Of course, the primary condition that affects the efficiency of any solar system is the amount of solar radiation that strikes an area for the particular period in which you need to use the solar heat. If you plan to install a space heating set-up, you should know the efficiency of the solar collector during December, January and February, when you will need the system most.

How much solar radiation can you expect in your area? That depends on the amount of sunshine and atmospheric conditions where you live. In some areas, there are so many cloudy days in winter that the amount of radiation reaching the collector will be too small to make it practical to install a solar system. In other areas, sunny days and chilly nights for many months of the year make solar an attractive proposition.

Solar insolation, or the amount of solar energy at any one place, is measured most often in *Langleys*. One Langley equals one gram-calorie per square centimeter per minute, and is measured with an instrument called a "pyranometer." The average amount of solar radiation reaching the earth's atmosphere per minute (called the "solar constant") is just under two Langleys. This is the equivalent of 442.4 BTUs per hour per square foot of area; or 1395 watts per square meter.

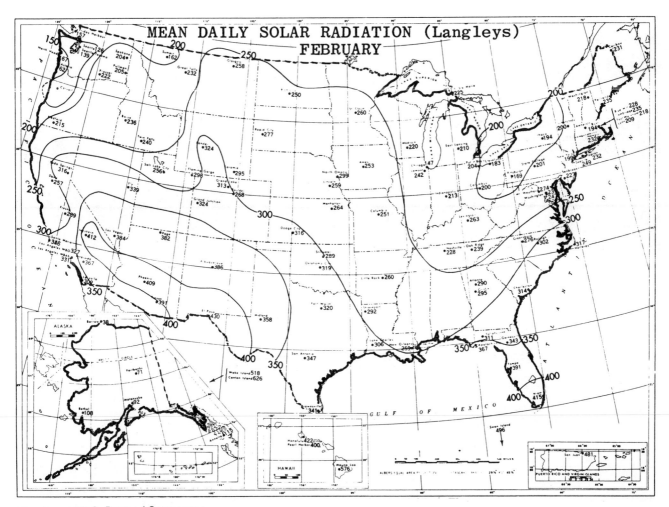

Courtesy of U.S. Dept. of Commerce

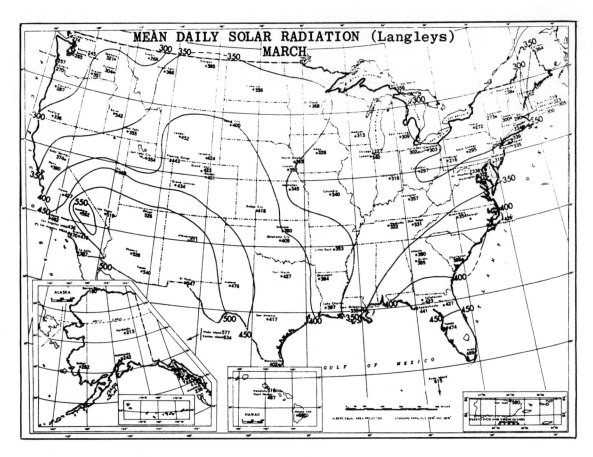

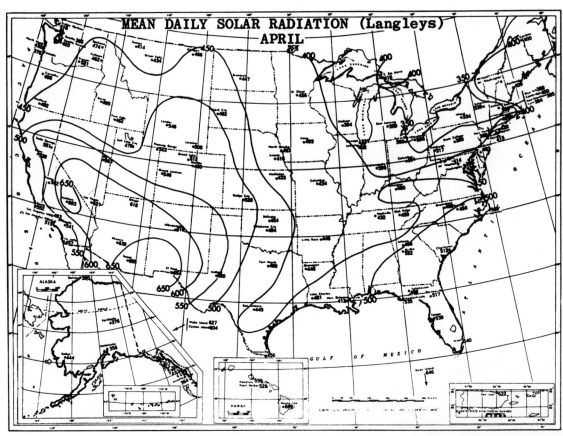

58

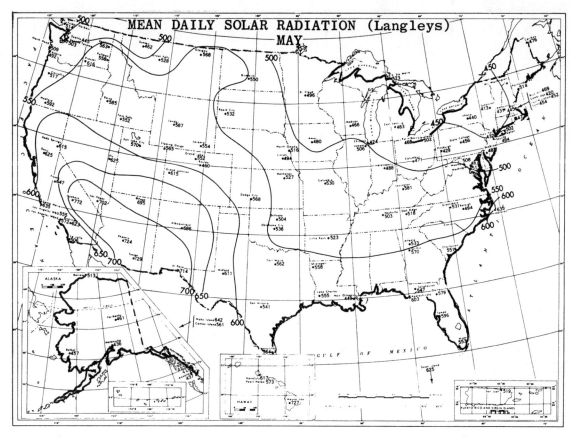

MEAN DAILY SOLAR RADIATION (Langleys)
MAY

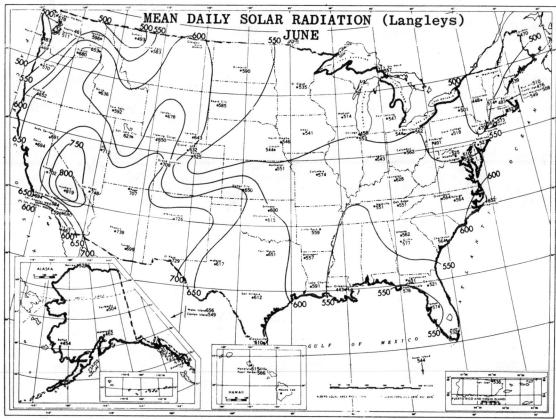

MEAN DAILY SOLAR RADIATION (Langleys)
JUNE

Courtesy of U.S. Dept. of Commerce

59

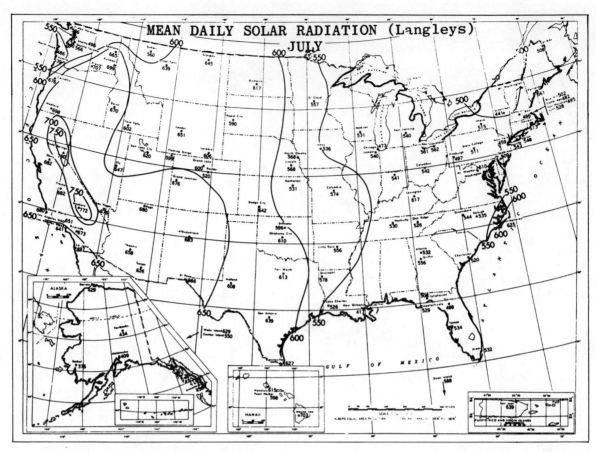

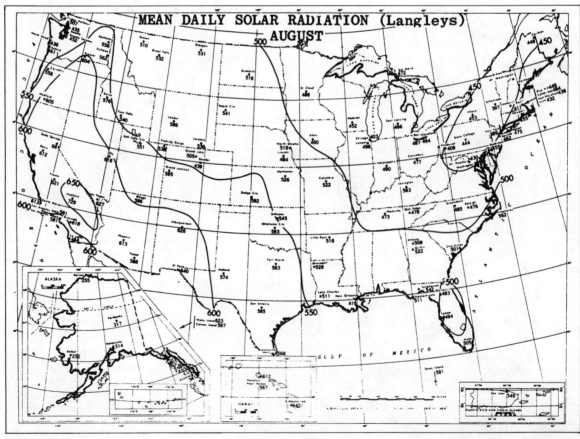

Courtesy of U.S. Dept. of Commerce

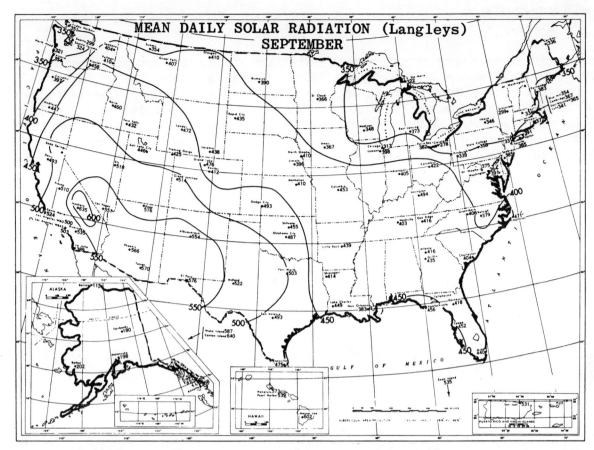

MEAN DAILY SOLAR RADIATION (Langleys)
SEPTEMBER

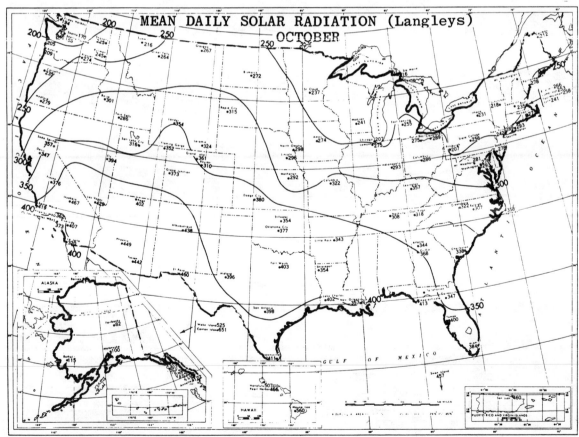

MEAN DAILY SOLAR RADIATION (Langleys)
OCTOBER

Courtesy of U.S. Dept. of Commerce

However, the amount of solar radiation received on the earth's surface at any point is affected by the density of the atmosphere and the angle from the sun. The National Oceanic and Atmospheric Administration (NOAA) measures solar radiation data daily at about 130 locations. The maps provided here show mean daily radiation by month. The U.S. Department of Commerce publishes a U.S. CLIMATIC ATLAS that gives weather data for many selected areas of the country.

By referring to the maps on these pages and doing some guesswork or interpolation for your area, you can come up with a rough estimate of the solar radiation. To put the information into useable form, you need to convert "gram-calories per square centimeter" to "BTUs per square foot." To find BTUs per square foot for solar radiation falling on the earth's surface, multiply the figure given in Langleys by 3.69.

(If you want to do your own arithmetic, one square foot equals 929 square centimeters; one gram-calorie is the amount of heat needed to raise the temperature of a gram of water by one degree Celsius at sea level. One BTU is the heat needed to raise the temperature of one *pound* of water by one degree F. at sea level, and one pound equals 435.59 grams. Don't forget to convert the Celsius temperatures to Fahrenheit.)

Let's work through an example. The mean daily solar radiation at Pittsburgh, Pa., in January is 94 Langleys. That's 347 BTUs per square foot per day at the earth's surface, or 10,757 BTUs per square foot for the 31 days of January. A 500 square foot solar collector at Pittsburgh with a designed efficiency of 50 percent should collect about 5,078,500 BTUs of heat in January. Suppose that collector is used to help heat a 2,000 square foot house in Pittsburgh that requires 37,000 BTUs of heat per degree

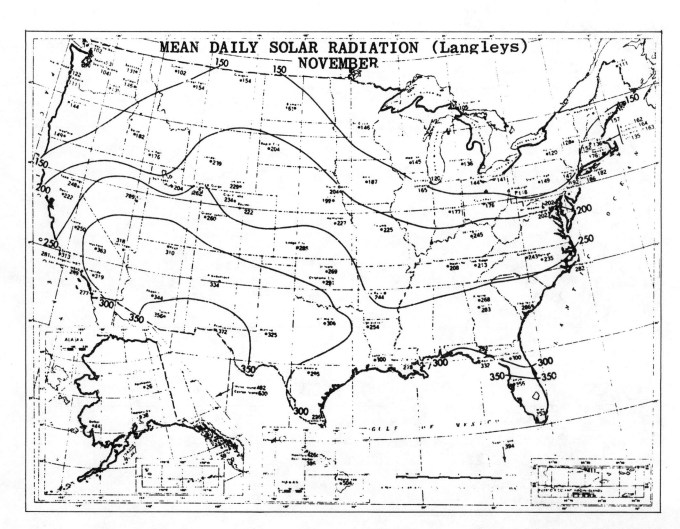

Courtesy of U.S. Dept. of Commerce

day. By referring to the heating degree day table in Chapter 3, we find that Pittsburgh has 1,119 degree days in January. That means a total heating requirement of about 41,403,000 BTUs for the month. In other words, for purposes of this example, the solar system would provide about 12½ percent of the heating needs in January.

If we go through the same exercise for a similar house in Albuquerque, N.M., we find that 305 Langleys per day fall to earth. That's 1,125 BTUs per square foot, so the 500 square foot collector (at 50 percent efficiency) would gather on the order of 8,721,850 BTUs of heat. Coupled with fewer degree days—930 compared with Pittsburgh's 1,119—the solar system provides more than 25 percent of the January heating needs in New Mexico.

There's more to figuring your heating needs than that, of course. For one thing, computing with accuracy the heat loss of a house is a fairly complicated undertaking. If

you need to get very precise about it, you should call on the services of a heating engineer.

Another way to estimate the BTUs of heat needed to heat your home is to total the petroleum or electricty consumed with your present heating system, then use the information on fuel efficiencies in Chapter 3 to compute the BTUs used.

The main point you can take from all this arithmetic is this: you can, in most areas of the country, build a solar heating system that will provide 50, 60 or 70 percent of your *annual* heating needs. However, the supplemental or "back-up" heating system will need to be scaled to heat your house during the time when you're getting only 10 or 20 percent of the total needs from your solar system. In the examples above, the solar heating system might supply 50 or 60 percent of the heat demand for April in Pittsburgh; and perhaps 80 to 90 percent of the April de-

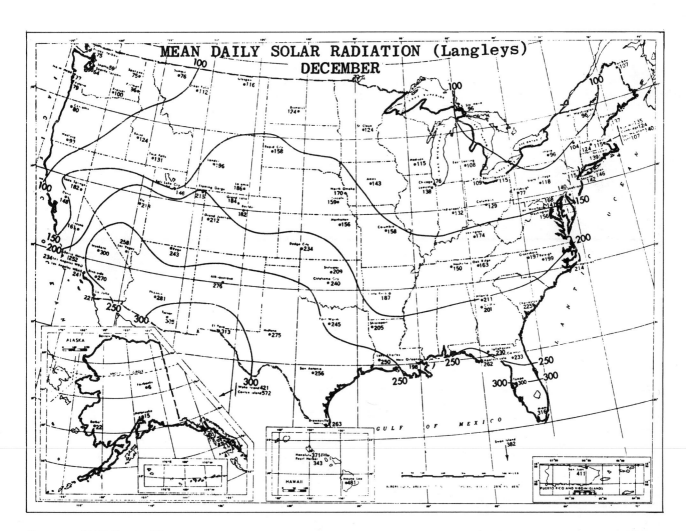

Courtesy of U.S. Dept. of Commerce

mand for the house in Albuquerque. But it's what the system does in January that dictates the supplemental or conventional heating source.

Which System: Active or Passive?

The state of the solar art, as far as home heating and domestic water heating are concerned, is focused on two broad categories: passive and active.

When most people think of solar energy systems, they usually visualize steeply pitched roofs covered with glass collectors. But there are other ways to capture the sun's energy: passive solar systems that make the house itself a solar heating unit.

Passive solar systems include the orientation of the building on the site, types of materials used, design of the system, good insulation and placement of collector surfaces in relation to the space or material to be heated. A passive solar heating system uses the actual structure of the house to collect and store heat from the sun.

On the surface, you might think that passive solar heating is less complicated than active systems. And they are, in a way. Passive solar requires no pumps, blowers or control systems to operate them.

But for an effective passive solar heating system, design is probably more important than with an active system. That's why designing and building a passive solar home can be even more challenging than installing an active system. With an active set-up, you always have the option of putting in another pump or blower. But if you have a poorly designed element in a passive solar system, the whole thing fails to work as it should.

Passive solar systems use gravity, heat flows, evaporation or other acts of Mother Nature to collect, move,

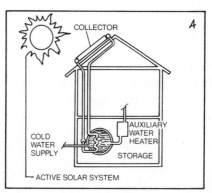

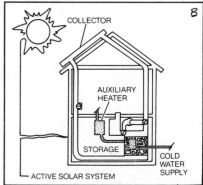

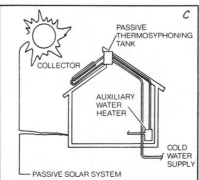

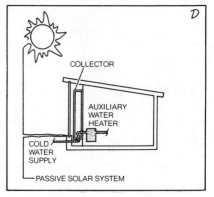

Solar hot water systems, designed to preheat water before it enters a conventional water heater, can be combined with a solar space-heating system or designed as a separate unit. Figures A and C show typical active and passive solar systems for preheating domestic hot water. Figures B and D show how a domestic water preheater can be combined with active and passive solar space-heating.

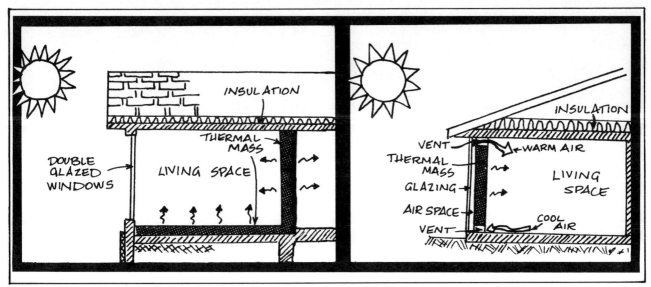

A direct-gain passive solar system stores heat in floors and walls of the living space, as illustrated at left. A thermal wall (also called "Trombe" wall) passive system utilizes a massive masonry wall for heat storage.

store and use heat energy without mechanical devices. The structure must fit the site and climate exactly, and all other elements—the floor plan, the house volume and position, choice of building materials—must be carefully geared to these features.

The simplest passive solar system is one that allows *direct gain* of solar energy through a large expanse of south facing double glazed glass. A rule of thumb says you'll need about 20 percent as much glass as you have liveable floor space in the house: a 2,000 square foot house would need about 400 square feet of windows facing south.

Heat from the sun warms the living space and its contents directly. In most designs, thick masonry walls or floors collect and store heat. In mild weather, heat can be stored in the masonry for about 12 hours; but only for 4 to 5 hours in colder climates.

Of course, all the glass not only lets heat in on sunny days, it can also lose a lot of heat at night and on cloudy days. That's why insulating shutters or curtains are vital to a direct-gain passive system. These also prevent overheating during the day. However, at best, the heat delivery with a direct-gain system is tough to control.

Another passive solar idea is a *thermal storage wall* (often called a "Trombe" wall, after its inventor, Felix Trombe). In this system, sunlight enters through large south-facing glass areas, then is converted to heat and absorbed by a massive masonry wall directly behind the glass. A narrow air space between the glazing and the masonry wall allows some convection heating of living spaces, in addition to the heat that radiates from the wall.

Again, with the Trombe wall, as with direct gain, some kind of movable insulation is needed to prevent heat loss to the outside at night, and to prevent excessive heat gain in warmer weather. Disadvantages with the thermal wall system include reduced visibility out of south windows, and the cost and structural support needed for the heavy masonry wall.

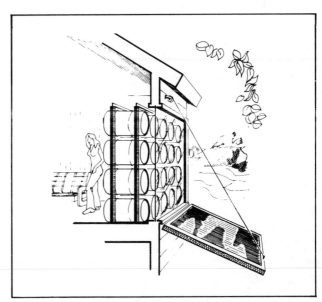

A "drum wall" solar system stores heat in containers of water behind the window glass area. The shutter panel has a reflective surface to direct more sunlight on the storage containers, and can be closed at night to prevent heat loss.
(Courtesy of National Solar Heating Center)

Still another approach is a modification of the Trombe wall to employ water stored in containers to collect and hold the heat. The use of 55 gallon drums filled with water was pioneered by Steve Baer, a New Mexico solar developer, and has been adopted by several solar builders.

Solar purists maintain that passive solar systems are the only type responsible builders should construct. For a practical matter, however, most solar systems utilize features of both passive and active techniques. These hybrids combine some of the economic values of passive design with the versatility of an active system with its pumps or blowers.

Active solar systems, well designed, can work in almost any area. If you want to build a solar heating system but don't have the proper roof orientation, you can even install a stand-alone solar unit in the backyard.

Initial costs for active systems are higher than for some passive systems, primarily because you need to buy more hardware. But costs may be more economical in terms of usable BTUs delivered per dollar invested.

Also, active systems are not as architecturally restraining as are passive systems; nor as demanding of attention while they are in operation.

Active systems are used to provide domestic hot water and space heating. Most active systems in homes or commercial buildings have four basic components: a collector to gather heat from sunlight; a storage system to accumulate the heat for use at night and on cloudy days; a delivery system to bring the heat to the spaces to be warmed; and a control system to automatically operate the system.

The most basic active solar system is a domestic water heater piped to a set of collectors; however, the solar system usually needs to be backed up with a conventional water heater.

The next step up is solar space heating, with the components designed into a complete system, or with solar heated air routed from collectors to the return air ducts of a conventional furnace. Or, solar heated water can be piped to coils in the return air ducts. Either way, the blower

Both air and water systems have advantages and drawbacks. Air is not subject to problems of corrosion and freezing, but is less efficient in gathering and transferring heat.
(Courtesy of U.S. Dept. of Energy)

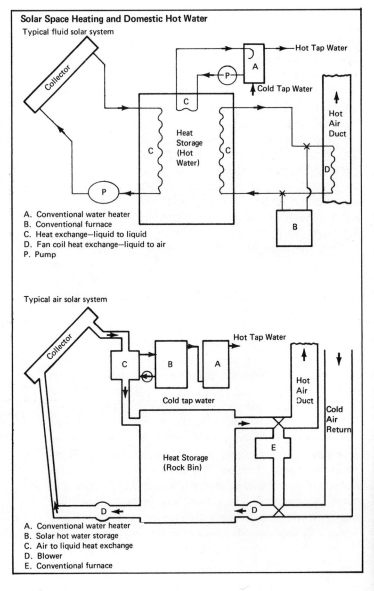

Solar Space Heating and Domestic Hot Water

Typical fluid solar system

A. Conventional water heater
B. Conventional furnace
C. Heat exchange—liquid to liquid
D. Fan coil heat exchange—liquid to air
P. Pump

Typical air solar system

A. Conventional water heater
B. Solar hot water storage
C. Air to liquid heat exchange
D. Blower
E. Conventional furnace

Solar attic house built by the University of Missouri engineers features 430 square feet of collector area that heats air in the attic. The heated air is circulated to living spaces, if needed, or to a rock storage bed beneath the house.

Solar water heating systems are usually water "preheaters" that warm up water before it goes to a conventional water heater.

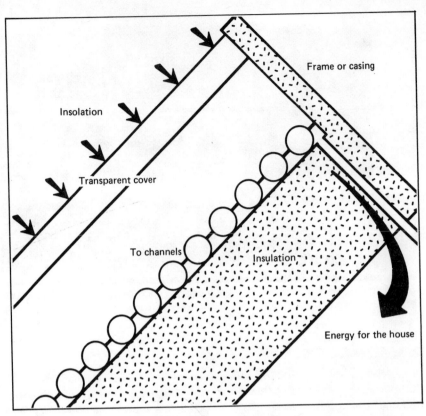

A flat-plate collector is essentially a glass covered box with a heat absorbing surface and insulation in the back. (U.S. Dept. of Energy)

Labels in figure: Insolation · Frame or casing · Transparent cover · To channels · Insulation · Energy for the house

on the conventional furnace distributes the solar heated air throughout the house.

From here, the options get more complicated.

Air, Water or Both?

Active solar systems use air or water—sometimes both —to transport the captured heat from the collectors to the living space or to storage containers. Both mediums have advantages and disadvantages.

Air collectors are less risky and simpler to build and operate than water systems—usually cheaper, too. But air is not as good a heat transfer medium as water or other liquids. As a result, the collectors are somewhat less efficient. So is the rest of the system.

When liquid is circulated through a collector, that medium is more efficient at transporting and holding heat than is air. However, with water, there's a danger of freezing in many parts of the country. Some method to prevent freezing must be used.

An antifreeze solution can be circulated through the collectors, but this cure presents some additional problems. For one thing, some sort of heat exchanger must be incorporated to transfer heat from the antifreeze to the water storage tank, and a heat exchanger loses some efficiency, because some energy is lost in the transfer process.

Most types of antifreeze are too costly to fill a 1,000 gallon storage tank (a common size for residential space heating systems). At any rate, this option is out for those systems that provide both domestic hot water and space heating. Antifreeze solutions are toxic or corrosive.

An alternative to using a primary circulating loop of antifreeze to prevent freezing is to build a drain-down provision into the system to prevent freeze damage on cold nights and cold, cloudy days. It isn't difficult to install a system that automatically drains down when the circulating pump shuts off on a temperature control switch.

Some hot water systems are designed to recirculate hot water from the storage tank through the collectors when danger of freezing exists. However, this seems like a costly solution—so is the now-and-then used technique of wrapping exposed pipes with electric heat tape.

COLLECTORS

A solar collector is a device to gather sunlight (radiant energy) and convert it to heat (thermal energy). Air, liquid or both is circulated through the collector to pick up the heat and carry it to its destination.

There are three basic types of collectors: flat plate, evacuated tube and concentrating collectors. Flat plate collectors are by far the most common type used, for both space heating and domestic hot water. Evacuated tube

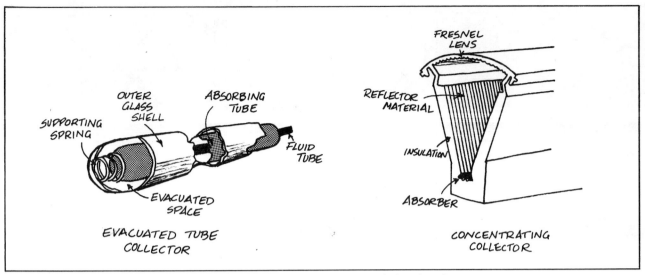

Concentrating solar collectors are used where high temperatures are needed, as in solar air-conditioning and some industrial processing.

and concentrating collectors are used mainly for high temperature systems such as solar air-conditioning or industrial processing.

For space heating, figure about 1 square foot of collector surface for every 3 to 4 square feet of house floor area. If you have a well built, well insulated 2,000 square foot home in a cold but sunny climate, you'll need about 700 square feet of collector to supply 60 to 70 percent of your annual heating requirements.

Dull surfaces absorb and radiate more heat than do bright surfaces; rough surfaces absorb and radiate more heat than do smooth ones. The working surface of a flat plate collector is an absorber plate made of metal or other material, and may have fluid tubing bonded to it or as an integral part of the plate. Heat transfer liquid passes through this tubing in water systems; in air systems, a stream of air is passed across the surface of the collector.

The plate is sealed in a frame. The back of the absorber plate is protected by a layer of insulation. The top or front surface of a flat plate collector is covered with one or two layers of glazing (glass or a special type of plastic, such as Tedlar). An air space is formed between the glazing and the absorber plate.

Collectors should be oriented due south for maximum efficiency. However, flat plate collectors can accept direct or reflected sunlight from a wide range of angles; the collector can be aimed from 15 degrees east of south to about 35 degrees west of south with little drop in performance. Thus, house design and orientation need not be locked into a true south position.

The angle, or "tilt", of a collector in relation to the sun's angle depends on location. If a solar system is designed for winter heating only, the collector surface should be nearly perpendicular to the sun's rays in the middle of January—generally the coldest time of the year. This right angle position is the sum of your latitude, plus 15 degrees. If you live at about 40 degrees North latitude, the collector should be angled at approximately 55 degrees from horizontal. However, the "tilt" can be off by as much as 10 degrees without having any noticeable effect on the collector's capability.

If your solar system is for both heating and cooling, the optimum angle is near the latitude angle.

A well designed flat plate collector can get to 200 degrees F. or hotter. But high-end temperatures are not as important as efficiency in a collector for space heating or domestic water heating.

"There's not much need to let the temperature go over about 150 degrees," says Jerry Newman, engineer with the U.S. Department of Agriculture at Clemson University in South Carolina. "The higher the temperature in the collector, the greater the temperature *difference* with outside air, and the more heat that is lost to the atmosphere. We try to design systems that collect the most BTUs, not collectors that attain the highest temperatures."

Newman and his colleagues at Clemson designed several of the solar homes pictured here. As Newman points out, BTUs and degrees of temperature are different creatures. For example, there are more BTUs in a bathtub of 70 degree water than in a teakettle of boiling water.

Evacuated tube collectors are built rather on the principle of the vacuum (Thermos) bottle. The absorber tube is surrounded by two transparent glass tubes separated by

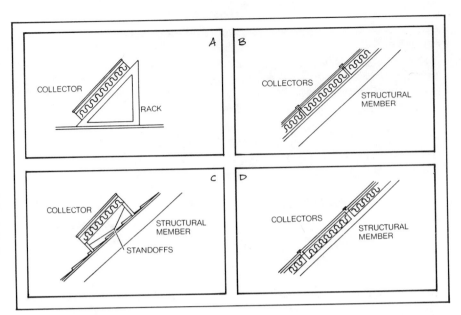

Methods of mounting flat-plate collectors on the roof: (A) mounted on a frame or rack, as on a flat roof or on the ground; (B) mounted directly on the roof surface; (C) mounted on stand-offs that separate the collector from the finished roof surface; and (D) built in as an integral part of the roof. (U.S. Dept. of Housing and Urban Development)

a vacuum. With no air to convect heat from the inner glazing to the outer glazing, this type of collector retains a much higher percentage of the heat absorbed than do flat plate designs. This makes the evacuated tube highly efficient, especially when the temperature differential between the absorber surface and outside air is great.

Evacuated tubes usually are installed in rows or banks, with reflective troughs beneath them. In systems that require temperatures of 200 degrees F. or higher, this type of tube is ideal. However, evacuated tubes are costly, and are not yet readily available in all areas.

Even higher temperatures are possible with concentrating-type collectors. If you have ever focused the sun's rays with a magnifying glass to burn a hole in a piece of paper, you have witnessed the principle behind the concentrating collector. These use special lenses or reflectors to multiply the amount of energy per unit of area striking the absorber. The absorber gets hotter quicker than with other types of collectors. It's not so much that more energy is collected as that it is collected on a smaller surface at a much higher temperature. In experiments, this type of collector has been used to focus enough heat to generate steam in a high pressure boiler.

Both evacuated tube and concentrating collectors are expensive—too expensive to use in most space heating or water heating set-ups. Also, both are less tolerant of indirect angles in relation to the sun than are flat plate collectors.

Collector efficiency relates to that fraction of incoming solar radiation captured by the collector. If a system gathers half of the incoming radiation, it is 50 percent efficient. Efficiency varies with outside temperature, whether skies are clear or cloudy, whether the wind is blowing, the design and insulation of the collector. No collector can be 100 percent efficient—to capture all the BTUs that fall on it. With good weather conditions, a collector that captures 55 percent of the available BTUs is fairly efficient.

Even homemade solar collectors are costly to construct and install; they should be built well enough to last for several years—ideally to endure as long as the building stands. Soundness of construction and materials used contribute to durability as well as to efficiency. The collector must be well sealed to keep out rain, wind and dust.

Now, let's look at the rest of the solar system.

STORAGE

There are three main methods for storing solar heat: water, rocks (or other masonry) and change-of-state storage (so-called "Eutectic" salts). Each has its own peculiar advantages and disadvantages.

Water has the advantages of low cost, availability and high heat capacity—or the ability to contain heat within a limited amount of space. However, water storage tanks, pipes and collectors must be protected from freezing where placed outside the heated space. Tanks also need

Solar-collected heat is commonly stored in either water (or some other liquid) or rocks. As a rule of thumb, two gallons of water storage are required for each square foot of collector area. With rock storage, about 0.5 cubic foot of rock is needed for each square foot of collector.

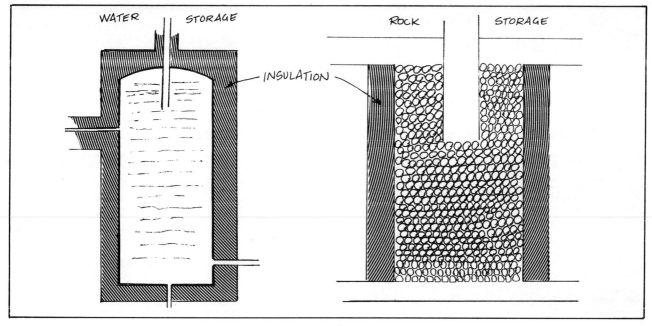

to be protected from rust and corrosion, either by using a corrosion inhibitor in the water or by building tanks of corrosion resistant materials.

Where corrosion inhibitors or antifreeze solutions are used, there's always some danger that these products can leak into the potable water system.

Plastic tanks are not subject to corrosion, but the kind of plastic needed to store large quantities of hot water is expensive, compared with the cost of steel tanks. Concrete is safe, durable and relatively economical non-corroding storage, but if a leak develops, it's tough to patch.

Another approach for some builders is to line the storage tank—of whatever material constructed—with replaceable bladders or diaphragms to insure water tightness and to protect the tank itself.

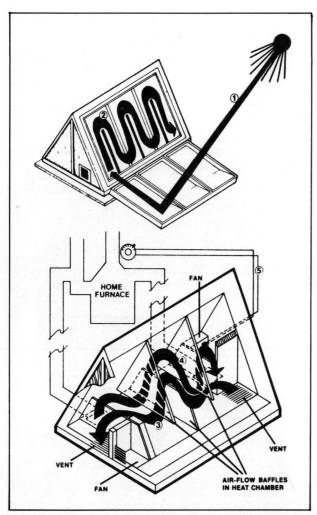

Stand-alone solar furnaces incorporate both collectors and rock storage in an A-frame structure. The unit is connected by insulated ductwork to the warm-air heating system in the dwelling. (Courtesy of International Solarsystems Corp.)

Weight is another factor. Two thousand gallons of water weigh about nine tons—plus the weight of the tank. Wherever the storage system is placed, adequate structural support is an important consideration. Storage tanks can be buried, either adjacent to the house or beneath the house, but must be protected with moisture proofing and ample insulation.

Despite its drawbacks, water is the most practical method for storage of heat with liquid systems at this stage of the game. Figure on needing one to two gallons of water storage for each square foot of collector surface.

For air type collectors, rocks or similar materials are the most convenient, cost effective way to store collected heat. Rocks are fairly available, economical and are not subject to some of the hazards associated with water—corrosion, freezing or leaking.

However, rocks do have a couple of drawbacks: one, they take up a lot of space for the amount of heat stored; two, they are much heavier than water. You'll need roughly three times as much space for rocks as you would need for water, for the same heat storage.

But rocks can be used to store heat at temperatures above 212 degrees F., if need be, while unpressurized water containers cannot carry temperatures above that point safely. However, that's a rather academic point with most home solar systems—temperatures usually don't go that high.

The best kind of rocks to use for storage should be rounded in shape and about two inches in diameter. Larger rocks offer less resistance to air flow, but do not have as much surface area for heat storage. Stones must be thoroughly cleaned before being place in the bin, and air going to the collector should be filtered.

Change of state storage materials are becoming more popular as different types of salts are developed. These substances change composition from solid to liquid when heated. The change allows the storage of more heat per pound than if the material did not change composition.

When these materials are cooled and go from a liquid to a solid state (at about 90 degrees F. for Glauber's salts), they give off the extra heat, in the same way that water gives up heat as it freezes into ice.

The one big benefit of change of state salts is that they can contain a great deal of heat in a limited space and at limited weight. Suppose you design a solar system that needs to hold 200,000 BTUs at 100 to 150 degrees F. Water systems would need 53 cubic feet at 3,300 pounds; rocks would require 175 cubic feet at 17,500 pounds; while Glauber's salts would require just 19 cubic feet at 1,740 pounds.

Eutectic salts are fairly reasonable in cost and widely available. However, packaging the material to use in a heat storage system can be expensive—more expensive than the material itself, in many cases. Another

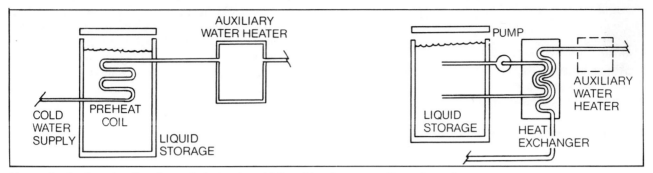

Two methods of pre-heating domestic hot water: at left, cold water passes through a coil immersed in the heat storage tank; at right, the heated liquid in storage is pumped through a heat exchanger or a conventional water heater.

drawback with most of the change of state materials used so far is that the salts can only go through so many cycles before they lose the ability to change state and automatically store and release heat.

SOLAR HOT WATER

If, for some reason, you aren't quite ready to go solar for space heating, you can still cut energy bills by installing a solar domestic water heating system.

These have a reasonably low first cost; water heating systems vary between $700 and $2,500, depending on the size and installation costs. A small solar heater installed by a homeowner would be at the lower end of the price range.

Solar water heaters typically have a shorter pay-back period than do space heating systems. One reason is that they operate year 'round, rather than just during the heating season, and domestic water heating can account for 20 percent or more of your heating bill. Most solar water heaters will repay their initial investment in 4 to 10 years, depending on cost and type of energy being replaced.

Water heating collectors can easily be retrofitted on existing houses. The collector area is small (usually less than 100 square feet), the storage tank small (80 to 150 gallons) and the capacity can easily be plumbed into existing conventional hot water systems.

The list of materials you'll need for solar domestic hot water systems is considerably shorter than needed for space heating. The simplest collector is an "open" system that circulates potable water through the collector. When the pump is not operating, the water automatically drains down into the storage tank to prevent freezing. The storage tank for a hot water system is often a conventional water heater tank (or tanks) without the heating unit, sized to meet the family needs. And, of course, you'll need the associated plumbing fixtures—pump, valves, temperature sensors to activate the system and circulate the water.

To size a system for your home, assume that each family member uses 20 gallons of hot water per day. A family of four would require about 80 gallons of hot water each day.

A solar water heater to provide 75 percent of that need would require a collector area equal to 1 square foot of area for each gallon of hot water per day, and a solar storage tank of about 80 gallons capacity.

Careful planning and sizing of the components of a solar heating system may let you step down a size or two in the electric motor required to operate pumps or blowers. (Courtesy of Grundfos Pumps)

With a "closed" water heating system, some kind of antifreeze would be needed, as discussed earlier. Here's how such a solar domestic water heater might work:

1. Cold water flows into the solar storage tank from the well or city water main.
2. A temperature differential sensor monitors both the temperature of the water in the tank and the temperature in the collector. When the temperature at the collector is 5 or 10 degrees F. higher than that at the tank, the sensor activates the pump to circulate an antifreeze solution through the collector.
3. The solution absorbs heat as it passes through the collector, then transfers heat to the potable water in the storage tank.
4. When hot water is needed in the house, it is drawn from the tank of the *conventional* hot water heater. The water heater tank is replenished with water from the solar storage tank.

This large swimming pool in Los Angeles is solar heated.

5. If the solar heated water is at the required temperature, the conventional water heater does not switch on. If the water from the solar storage tank is cooler than the setting on the water heater, the conventional heater comes on and heats the water as needed—but it only has to add a few degrees of heat to the water in most cases.

With most solar water heating systems, you'll need some safety devices to prevent scalding. Water at 190 to 200 degrees can literally cook human flesh. Many hot water systems use automatic mixing valves at sinks, lavatories and bathtubs.

Solar systems for heating swimming pools usually operate on the same basic principles as other solar hot water set ups, except that the storage area is the pool itself, rather than a hot water tank. One of the more effective—and economical—ways to heat a swimming pool is by using the pool itself as a collector, merely by

Solar Heated Swimming Pool
Los Angeles, California

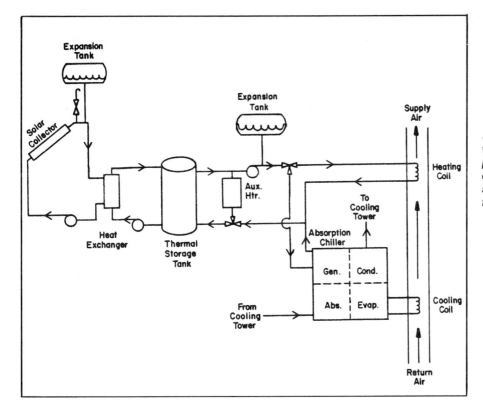

Solar air-conditioning systems are workable—much on the same principle as old gas refrigerators operated—but are priced out of sized dwelling would cost $12,000 to $15,000.

placing an easily removable transparent plastic covering over the water surface. In most areas, by using this covering collector when the pool is not being used, water temperatures can be kept above 80 degrees for several months without using other heat sources.

SOLAR COOLING

The heat from a solar collector can be used to power an air conditioning system, but the cost for such equipment is beyond the reach of most homeowners—starting at $15,000, for an average sized house.

Several solar air conditioning systems are possible, but most practical is the absorption refrigeration method —similar to oldtime gas-powered refrigerators. A conventional air conditioner compresses a vapor (usually the inert gas, Freon) by mechanical energy. The vapor is cooled and condensed into a liquid, then vaporized again into a chamber of lower pressure. As the liquid vaporizes, it withdraws heat from the space to be cooled.

A solar air conditioner pressurizes the refrigerant vapor by heating, rather than by mechanical energy. The vaporized refrigerant is recovered for recycling by absorption in a salt solution. The cooler air that results from evaporation of the refrigerant is ducted into the house; the warmer air from inside the house is dumped outdoors.

Other solar air conditioning systems, such as using a solar powered Rankine turbine engine to operate the compressor, are in the experimental stage. Early work indicates these systems are successful, but they would be prohibitively expensive for any kind of commercial use right now.

A more practical approach to space cooling is to design the solar heating system to draw in cool night air and store it in the water or rock storage for use to cool the house during the day. This is less effective than solar heating, primarily because the temperature differentials are small.

For example, when the outside temperature is 10 degrees F., a moderately efficient flat plate collector can pick up 120 to 150 degrees or more from the sun. That's a temperature difference of well over 100 degrees. In the cooling mode, nighttime temperatures in summer may only get down to 70 degrees; while temperatures the next day climb to 90 degrees or more. That 70 degrees "collected" during the night will not go very far toward cooling the house—not if it is used directly. But if that stored "coolness" is used in a heat pump's heat exchanger the heat pump can operate more efficiently. We'll have more about teaming up solar and heat pumps shortly.

PUTTING IT ALL TOGETHER

When designing any heating system, keep in mind these natural physical laws: heat always flows spontaneously from a warm surface or area to a cooler surface or area. Also, heat rises. Rock in a storage bin or water in a tank will be hotter at the top of the structure. You can design your system to take advantage of this stratification.

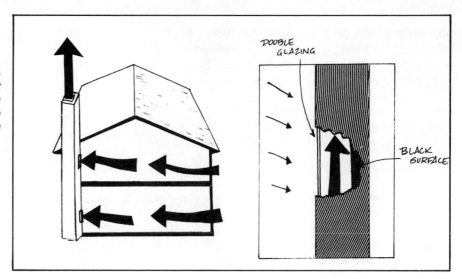

A solar chimney, a glass-fronted vertical duct, sets up a natural draft that can vent warm air out of living spaces and pull cooler air into the house through windows on the shady north side.

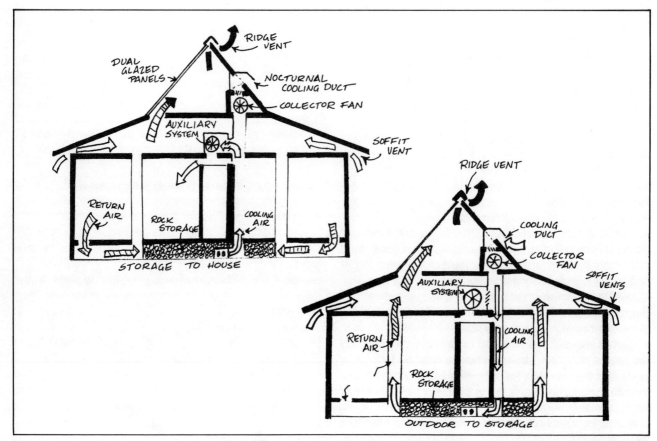

An active solar air system can store some nighttime "cool" for use when temperatures climb the next day.

By extracting heat from the top of a tank, for instance, you can get more work out of any heating system.

Planning the entire system on paper will prevent possible errors when the real work gets underway. Keep these goals in mind as you lay out the system:

1. The collector needs a good location, out of shade in winter (and at least partially in the sun in summer, for water heaters), facing south and tilted at the proper angle—latitude plus about 10 degrees of arc.

2. Pipe or duct runs between the collectors and storage or existing water heater should be planned to keep interior carpentry to a minimum. Also, lay out the plumbing with as few joints as possible—joints are subject to leaks.

3. Hold uninsulated pipe and duct runs to a minimum to cut down on heat loss.

4. Planning for the components of a solar system to be retrofitted should take into account existing plumbing fixtures and structural additions that may be needed.

5. Circulation pumps (for water systems) and blowers (for air) should be compatible with the rest of the set up. For each 40 square feet of collector area, a pump should circulate about a gallon of liquid; a blower should push 80 cubic feet of air per minute.

As a practical matter, some kind of domestic water heating device can be included in most solar space heating systems. In some installations, a heat exchanger leading from a conventional hot water tank is immersed in the solar water storage tank. In others, the heat exchanger is installed in the domestic hot water tank itself.

With air systems, the heat exchanger is placed in the path of the hot air coming from the collector. In some situations, the hot water tank itself is used as a heat exchanger and is positioned in the hot air flow, or buried in the rock storage pile to absorb heat.

Such combined systems can reasonably be expected to provide 40 to 60 percent of space heat and 75 to 85 percent of the hot water for a family of four living in a 1,500 square foot house. As noted earlier, the real benefit of many domestic water heating systems is to preheat the water so the conventional water heater doesn't have to work as hard.

Technically, it's possible to provide all space and hot water needs with solar heat. And in the southwestern U.S. where the sun shines 300 or more days per year, some homeowners are coming close to using solar 100 percent. But in most parts of the country, the amount of collector area and heat storage space needed for a total solar system would be impractical—so would the cost. A better answer is a back-up system of some kind.

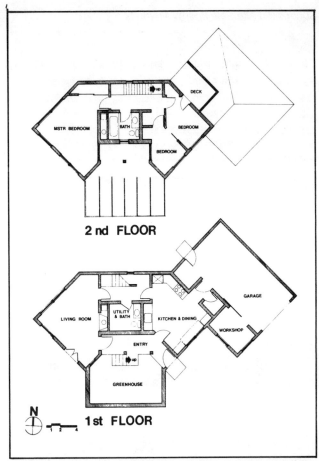

Floor plan of the "hybrid" solar home built in Utah by Robert and Lola Redford shows position of the solar greenhouse with living spaces arranged around the solar structure. The home was designed for the Redfords' ranch manager by architect Claire Pollack. (Drawing by Terry Rainey)

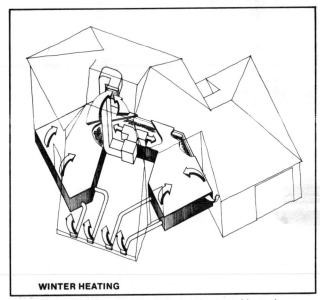

WINTER HEATING

Solar heated air from the greenhouse is stored in rock beneath main-level floors in the Redford ranch manager's house. Back-up heating systems are a heat-circulating fireplace and an electric resistance heater. (Drawing by Terry Rainey)

If you're adding solar to an existing system, you no doubt already have that back up capacity. Most common reserve heating sources are forced air furnaces and electric resistance heaters.

But heat pumps are gaining wider acceptance all the time, either as an in-between auxiliary system or as the only back-up to solar. If you're designing a solar system for new construction, you may want to give some thought to a marriage of the two.

A heat pump is basically a reversing refrigeration unit. It can pump heat from the indoors for cooling, just as an air conditioner does. But it can also reverse the cycle to extract heat from outdoor air to heat the indoors.

The problem is, as the outside temperature drops, so does the heat pump's efficiency. At the same time, colder temperatures make space heating requirements go up. In most conventional heating systems, the heat pump is supplemented at these times by electric resistance heating coils.

In some newer installations, however, heat pumps are being used to draw heat from solar heated air, rather than from the cold outside air, to increase the efficiency of both the heat pump and the solar installation. The heat exchangers from which the heat pump draws its BTUs can be connected to either a solar air or water system.

In such systems, the heat which the heat pump delivers to your house is the sum of the heat extracted from stored solar heat plus the electric energy needed by the compressor of the machine. The efficiency of a heat pump (called the coefficient of performance, or COP) is measured by dividing the total BTUs delivered by the BTUs supplied in the electricity that runs the pump.

In a conventional set up, a COP of 1.5 to 2 is fairly typical. In other words, the heat pump delivers 1.5 to 2 BTUs of heat energy for each BTU of electrical energy used. By combining a heat pump with a solar system, the COP of the heat pump can be boosted to 3 or more.

A key advantage to using solar in combination with a heat pump is that the solar system can be designed to operate more efficiently. As we mentioned earlier, the efficiency of a solar system increases as the temperature of the liquid or air pumped through the collector decreases —there is less temperature difference with outside air, thus less heat lost from the system. A properly designed heat pump solar energy system will have the collectors working at peak efficiency and the heat pump supplying only the additional heat required.

The system also works more efficiently in summer. Systems using a heat pump (rather than a Freon air conditioning system) for summer cooling can use the solar storage to pile up nocturnal "coolness", then use this cooler air or water in conjunction with the heat pump to cool the house. The heat pump is working with a temperature of, say, 75 degrees, rather than with outside air at 95

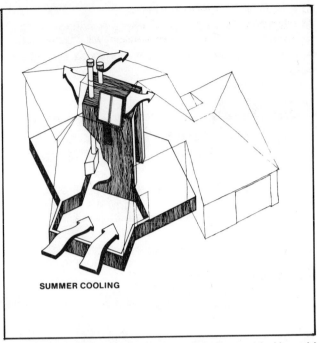

SUMMER COOLING

Summer cooling of the Redford house is provided by cold spring water circulated through coils in the central air duct; the same ½ h.p. blower that circulates heated air in winter pushes cool air through the house in summer.
(Drawing by Terry Rainey)

degrees or more. This kind of combination may save you even more money if your local utility offers a reduced "off-peak" pricing schedule—and more companies are going to this kind of incentive all the time. Generating costs are higher per kilowatt/hour during periods of high use; lower during periods of low use.

IN SEARCH OF THE QUALIFIED CONTRACTOR

Don't be too surprised if your telephone directory's Yellow Pages doesn't have anyone listed under the heading "Solar Expert." The solar energy industry is still a toddler, and there are many unproven businesses and workmen involved in it—at last count some 3,000 companies are building solar equipment. A vital factor in the performance of a solar system—as with house construction in general—is the quality and durability of materials and workmanship.

Your local association of home builders is a good place to start looking for a solar contractor. Some utility companies also work with experienced contractors. A list of solar builders and contractors in your state is available from the National Solar Heating and Cooling Information Center, P.O. Box 1607, Rockville, Md. 20850. You might also want to contact the American Institute of Architects, at 1735 New York Avenue, N.W., Washington, D.C. 20006, or the American Society of Heating, Refrigerating and Air Conditioning Engineers, 345 East 47th Street, New York, N.Y. 10017.

Poor quality of work is the number one cause of solar energy system problems. Unfortunately, not every contractor understands solar energy and how it works. When buying solar equipment and hiring someone to install it, use the same judgement as when making any major investment.

When contacting any supplier or installer, find out his performance record. How many systems has he installed? Do members of the firm have heating, plumbing and engineering experience and credentials? Most states require that architectural and engineering firms be licensed. Make sure the solar professional you hire is properly accredited in your state.

Each collector's manufacturer should have published information on the BTUs his unit produces per square foot per heating season month. About 10,000 BTUs per square foot per month is theoretically available at about 40 degrees north latitude, but 3,000 to 3,500 may be closer to reality. If the salesman cannot give you the kind of information you need, go to another dealer.

Ask questions of contractors, dealers and installers of solar equipment, and get the answers in writing:

1. Is the unit reliable and durable? You are buying a system that should last 20 years or more with only minor maintenance and repairs; examine the materials and construction methods until you're sure that you'll get your money's worth.
2. Examine the company's past record, if possible. Check out the warranties offered to cover hardware and installation work. If something goes wrong, who will fix it? How long will it take to repair and are repair parts easily obtainable? Who pays for what and for how long?
3. Is the contractor familiar with local plumbing and zoning codes? Better yet, check local codes yourself before buying or installing a system.
4. Solar collectors should be protected from both overheating and (in the case of water systems) from freezing. What provisions are made for these safety features? There are workable solutions for both potential problems.
5. Pipes, ducts and backs of collectors—as well as storage tanks and bins—must be insulated to prevent heat loss. What kind of insulation does your contractor recommend? Some types are better to use than others.

For example, some foam insulation starts to vaporize at about 170 degrees F. and should not be used to insulate any components that reach high temperatures.
6. Make sure the system is taken through a complete operation test before you accept it—or pay for it. All controls should work and any leaks should be repaired immediately. Insist that the system be test run long enough to confirm that the collector does indeed pick up solar energy and that it does increase the temperature in the storage.
7. Make sure the installer leaves you an operating manual for your system, with detailed instructions on how to isolate sections of the system if problems—such as leaks—occur.

Costs

Costs for a typical (if there is any such thing as "typical") solar set up for both space heating and domestic hot water will run from $2,500 to $12,000—sometimes more.

That's quite a price spread, but the top end figure is based on buying all factory made components and hiring all construction and installation work. A homeowner who is skilled in carpentry, plumbing and electrical work, and who shops around for materials, can bring in a solar system well toward the low end of the range.

A big part of the cost is tied up in collectors. Flat plate collectors ready made can run from $5 to $20 per square foot of collector surface. If you build your own, they don't have to be nearly that expensive to be workable. The 4x8 foot collectors built by J.W. Fish, and described in Chapter 2, cost just over $65 each to build—not counting any charge for Fish's time and labor. That's a cost of about $2 per square foot of collector surface, and if the collectors on Fish's roof are marginally less efficient than prefabricated ones might be, the difference in cost will pay for some loss of performance.

Two important factors in evaluating any solar system are initial cost and system efficiency. A relatively inexpensive collector with low efficiency may be a poor choice compared with a more expensive one that captures and delivers the sun's energy more efficiently. All other things being equal, the solar system that delivers the most heat per dollar is the best buy.

Other financial considerations may not be evident at first. Investigate well ahead of time for possible property tax exemptions, tax credits and sales tax exemptions and rebates. Keep in mind also that "Uncle Sam" will help pay for a solar installation.

All greenhouses are solar structures, but many attached greenhouses have design-in features that make them efficient collectors of solar heat for the adjoining living spaces. (Courtesy of Lord & Burnham)

5
SOLAR GREENHOUSES: A GROWING IDEA

"We can store enough energy in winter to keep temperatures above 65 degrees F. for two days or longer."

Odell Morgan
Solar greenhouse owner
Norman, Oklahoma

We might well have included attached greenhouses as one kind of solar heat collector in Chapter 4.

After all, *any* greenhouse is a solar structure, to some extent. And you probably have noticed that several solar designs incorporate an attached greenhouse as part of the energy package: the Redford ranch manager house and the University of Missouri solar house, for examples.

But there are enough unique characteristics and benefits of a well planned solar greenhouse to warrant devoting a separate section to this energy saver, particularly as a retrofit consideration. Here are some of them:

• An attached greenhouse can provide a good deal of the heat needed in the living space, depending on size and design.

• It's an inexpensive way to gain experience in collecting and using solar energy; much less costly than most of the solar systems outlined in Chapter 4.

• It's an economical way to add livable space to a house. For an investment of $2 to $8 per square foot, a greenhouse can serve as a breakfast nook or a "garden" dining room.

• And, naturally, a greenhouse provides a controlled environment for growing off season ornamentals and vegetables.

A solar greenhouse attached to a south wall of a residence *can* be all of those things. Unfortunately, it cannot be all of them equally well at the same time.

If your main objective is to collect and store heat for space and water heating, the operation of an attached greenhouse for these purposes often will be at odds with the steady temperatures and humidity levels demanded by many species of plants.

For instance, allowing the greenhouse to collect the maximum amount of heat on sunny days may push temperatures inside the structure too high for some plants. The procedure in a conventional greenhouse would be to vent this excess heat off to the atmosphere. However, if your goal is to gather as much heat as possible, you aren't likely to want to waste any of it.

Also, a conventional greenhouse would be heated from an auxiliary source to keep temperatures above a certain minimum at night and for prolonged cloudy periods. If the goal is energy conservation, an attached solar greenhouse might be allowed to cool down during these times—rather than make a drain on space heating sources—perhaps below the low-end temperature needed to grow many plants successfully.

Because of these conflicts, many owners of attached greenhouses use their structures in winter as daytime heat collectors, rather than as a place to raise crops. But, this does *not* mean you cannot grow flowers and food if you choose those plants known to be hardy.

We are primarily interested, in this book, in saving and producing energy for heat and power. The features emphasized most are those that equip a greenhouse as a solar heater. Also, your greenhouse will have to be designed principally as a solar energy collector if it is to qualify for tax credits under Internal Revenue Service rules. The government is not yet interested in allowing energy

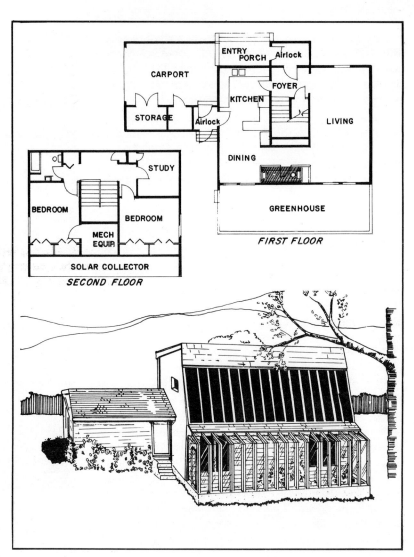

Several solar home designs employ both greenhouse and solar collectors, as in this residence plan by the Rural Housing Research Unit of Clemson University.

tax credits to homeowners for a structure that results in a net heat *loss*, as most conventional greenhouses do.

ENERGY COLLECTION AND STORAGE

The principles of collecting and storing solar energy with an attached greenhouse are virtually the same as with any other kind of solar device, as described in the previous chapter. The exercises for computing the available solar energy are also the same; as are the methods of collecting, transporting and storing heat—active and passive.

In a solar greenhouse, a double layer of glass or clear plastic lets the sunlight pass through. Heat-collecting surfaces behind the glazing absorb the light and convert it to heat. The greenhouse itself is a solar *collector*, and if well-designed, can be a very efficient one.

Generally, the idea is to collect enough surplus heat on sunny days to supplement residential space and/or water heating, as well as store up enough reserve heat to at least make the greenhouse itself self-sustaining at night and during cloudy periods. This is different from a conventional greenhouse. Many solar greenhouse designs have glazing only on the south wall—or perhaps the south wall and roof—and have the other three walls well insulated. Often, the interior surfaces of solid walls are covered with reflective material to provide more light for plants and the heat collector.

Insulation of the unglazed wall and roof surfaces is vital; as is insulation of the heat storage unit. Some builders install moveable shutters or curtains to provide nighttime insulation over the glass surfaces as well. Others pump foam "beads" into the air space between the layers of glazing for insulation at night, then pump the insulation back out to collect sunshine in the daytime.

Heat typically is stored in rock (air systems) or water, as with other solar heating systems. With some active solar greenhouses, the heat is pumped or blown to storage inside the living spaces. However, most solar greenhouses incorporate heat storage in the facility itself.

Attached greenhouses are most often built as passive systems. Water in plastic milk jugs, wine bottles, beer cans, 55-gallon drums or other containers stores the heat collected. The principles of an active solar system frequently are installed to move the stored heat into the living space or circulate heated water to the domestic hot water system.

How Big?

You don't need a lot of greenhouse floor space to cut heating bills significantly. If you are attaching a greenhouse to an existing dwelling, the size and shape of the greenhouse may be dictated by the length of a south facing wall.

Odell Morgan, of Norman, Oklahoma, built a 24x36 foot greenhouse attached to the south side of his home, right off the kitchen-dining room area. This is larger than most add-on solar greenhouses, but Mrs. Morgan does a brisk sideline business in plants, and the family converted part of the greenhouse to casual living space.

Morgan collects heat in black pipe snaked through the "attic" area of the gable roofed greenhouse, then uses a heat exchanger to heat water for storage in 55-gallon drums along both sides of the building. Because growing plants is of equal importance with space heating in the Morgan greenhouse, the system is designed with an auxiliary gas heater. However, the gas heater is seldom needed, except when the weather is cold and cloudy for more than three consecutive days.

One rule of thumb calls for an attached solar greenhouse to be roughly twice as long (east to west) as it is wide (north to south). This makes the ratio of collector surface to floor area about right for efficient energy gathering. For a practical matter, though, a greenhouse should be at least eight feet wide, whatever the length. Anything less than that will not have enough space to be useful as a plant growing area.

Generally, greenhouse additions built primarily for space heating range from 100 to 200 square feet, and will cost $3 to $6 per square foot for a handy do-it-yourselfer. If a contractor builds the structure, figure $8 to $12 per square foot for a passive structure. "Active" hardware is not included in these cost estimates.

What Type Construction?

The construction of an attached solar greenhouse should be a fairly simple project for a homeowner with

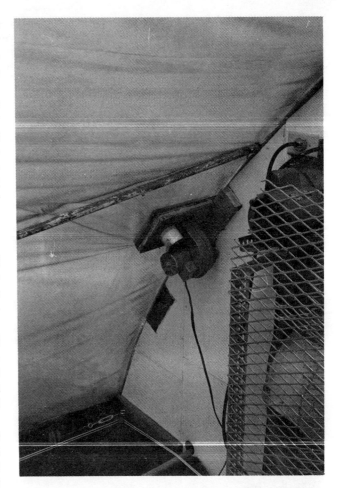

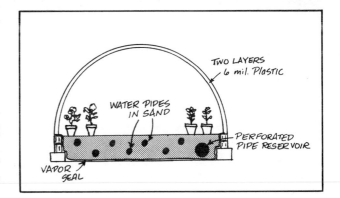

Quonset-type greenhouse at Oklahoma State University features detached solar collectors, which heat water that is circulated through pipes in the sand floor of the greenhouse. A small electric blower is used to keep a positive air pressure between layers of polyethylene film that forms the skin of the greenhouse.

This 8½-by-12-foot attached greenhouse at the home of Dewane and Maxine Thornton has a heat storage bin where broken bricks hold the collected solar heat. Thorton installed a 12-inch round metal duct at the highest point in the greenhouse to collect warm air. Two opposing fans are mounted on the storage bin—one to pull hot air from the collector through the brick; the other to push hot air from storage into the greenhouse.

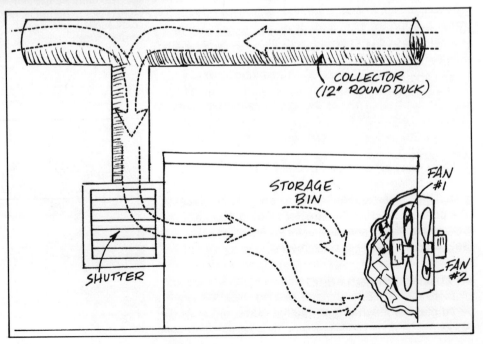

Larger than most, this 24-by-36 attached greenhouse built by Odell Morgan, Norman, Oklahoma, features a solar hot water system.

rudimentary building skills. However, unless you also have some experience in plumbing and electrical work, you may want to hire some help with the piping, pumps, blowers, controllers and other hardware.

The greenhouse should be located on the south side of the dwelling, of course. And it should be supported by a solid footing and foundation. From there, the size and shape of the space you enclose will be important considerations, as will your building budget.

What design appeals to you most? Or goes best with your house? Your attached greenhouse can have a vertical south wall. Or the south wall can be angled at about any degree of arc from vertical to a degree of tilt equal to your latitude plus 35 degrees. You may glaze only the south wall, the south wall and roof or south wall and at least part of the east wall, to pick up early morning sunlight.

How about materials? Do you want to build a permanent greenhouse that costs more money, or spend less money right now and experiment awhile? These are all decisions that only you can make, but here are some benefits and drawbacks with different materials and methods of construction:

Glass with metal frame. Glass holds its light transmitting ability longer than most plastic glazing materials; up to 90 percent light transmission for the life of the glass, if it is kept clean. This type of construction is usually more expensive than other methods. Also, if your region is subject to hailstorms or small boys with rocks, glass can suffer more damage than tougher glazing materials. Tempered glass is tougher than regular glass, but is more costly.

Control panel automatically operates pumps that circulate hot water from collectors to a heat exchanger in the center of Morgan's greenhouse; and also operates blowers at one end of the structure.

Solar plastics with metal frame. Plastics, such as Tedlar®, with good light transmission stand up well; some are guaranteed for 20 years. Construction costs can be fairly high; close to the cost of glass in many cases. Maintenance of high impact plastics is lower than with glass greenhouses. Poorer quality plastics degrade and "yellow" after awhile in sunlight.

Corrugated fiberglass on wood frame. If the roof is held with cables or clamps, rather than screwed or nailed, this type of construction is leak and insect proof. Construction costs are fairly reasonable. Less expensive fiberglass can "yellow" and cut light transmission in a few years, however, and untreated wood is subject to decay in a greenhouse environment.

"Half-Quonset" type fiberglass on steel. This method of construction is not often used for attached greenhouses, and not everyone likes the appearance of this type of structure. However, with arches every four feet or so, this kind of construction is easy to frame and fairly economical to build; although there usually is some waste of materials in cutting and fitting end walls to the rounded roof.

Polyethylene film on wood frame. This is the cheapest and least durable way to build an attached greenhouse. Sunlight and high winds wreak havoc with plastic films, and the covering probably will need to be replaced every year or two. With some film construction, a small blower is needed to keep a positive air pressure in the air space between layers of polyethylene. However, this method of construction can let you gain some greenhouse experience without a considerable outlay in building costs.

How Much Heat Can You Collect?

Not many attached greenhouses will heat an entire house—even in sunny weather. But most can provide 50 to 75 percent of the annual heat needed in a dwelling space 1.5 to 2 times the size of the greenhouse itself.

In other words, a 200 square foot greenhouse should provide most of the heat for 300 to 400 square feet of living space immediately adjoining the greenhouse. Or, with an active system, the same size greenhouse might provide 25 percent of the heat required by a 1,200 square foot dwelling.

The Griffin family of Santa Fe, N.M., kept records on

Collector pump circulates water through black plastic pipe in the greenhouse attic, then back to a large insulated tank in the center of the building. A second small pump circulates heated water from the tank through coils of pipe in 55-gallon drums filled with water—Morgan's heat storage system.

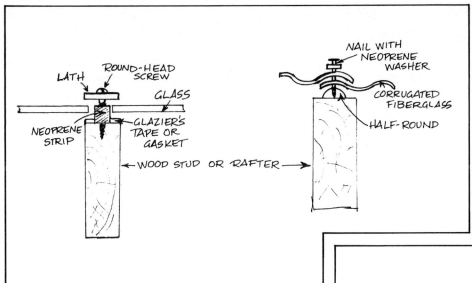

Glazing with different materials is handled in different ways. At left is one method of installing glass that allows for expansion and contraction with temperature changes (sections of glass should not be butted together; nor should sections of rigid plastic). At right is a fast, weathertight way to install corrugated fiberglass.

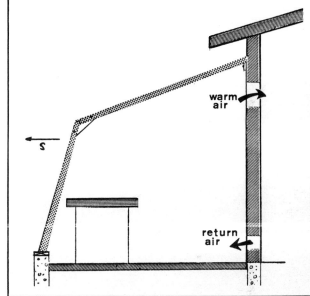

Add-on greenhouses ducted as above to the attached living space will provide heat to the rooms by natural convection.

the heating requirements of their 1,115 square foot home after adding a 120 square foot attached greenhouse, and note that the greenhouse cut their heating bills by 24 percent. The New Mexico Energy Resources Board's Solar Sustenance Project, for which the Griffins kept the records, estimates that such an attached greenhouse—in New Mexico—could provide 82,212 BTUs per day of surplus heat for home heating in fall; 151,464 BTUs per day in winter; and 312,702 BTUs per day in spring.

However, as mentioned above, the way an attached solar greenhouse is operated is the key to how much home heating help you get from it. Shutters, insulating curtains or some other heat stopper used at night and during cloudy weather can cut heat losses tremendously. Otherwise, much of the heat gained during the day can be lost through the glazing at night.

Many passive solar greenhouses retrofitted to existing dwellings are built so that doors or windows can be used as inlets for the collected heat. However, a more efficient system of performing this function is to build ducts—with closeable insulated dampers—at the bottom and top of the attached house wall. This way, the solar heated air from the greenhouse enters the upper duct by natural convection, while cool air from near the floor inside the house is exhausted into the greenhouse.

A greenhouse attached to a walk out basement or first floor wall might utilize a similar natural "thermosyphon" circulation to move solar heated water to a storage tank on the floor above.

Of course, mechanizing these functions can make the solar heated air or water go still further, with a small investment in electricity, and lets you turn more of the operation of the greenhouse over to automatic controls.

One final note on solar greenhouses: don't forget to provide ample ventilation for the structure for those seasons when that sun heated air is not needed. With some planning, ventilating shutters placed low in the east wall and larger exhaust vents high up on the west wall can take care of most greenhouse ventilation without the use of fans.

A pound of totally dry wood of any species has about 8,600 BTU's of potential heat.

6
HEATING WITH WOOD WITHOUT GETTING BURNED

"If one has cut, split, hauled and piled his own good oak, and let his mind work the while, he will remember much about where the heat comes from, and with a wealth of detail denied to those who spend the weekend in town astride a radiator."

Aldo Leopold
A Sand County Almanac

Some 500,000 Americans will install wood burning heating devices of one kind or another this year, making a giant leap toward the 1940's. As prices for natural gas, oil and other fossil fuels rise, more and more families warm to the idea of using America's "first fuel" as a plentiful, accessible and renewable source of heat.

There are several good reasons why wood is among the better natural sources of heat we have right now—at least for those who live near an accessible supply of wood. For one thing, our forests produce a surplus of wood, and probably will for some time to come.

"We face major demands on our timber; timber consumption in the United States may double in the next 50 years," Bob Bergland, secretary of the U.S. Department of Agriculture, reminded the Society of American Foresters, in October, 1979. "But our forests now produce only half of the wood they can grow naturally—perhaps only a third of the potential under intensive management."

In fact, selective cutting and improved management is needed right now in most forests and woodlands, to clear out diseased, weak and poorly shaped trees that interfere with the growth of better timber. Nationally, we are far from the point that over-cutting endangers forest lands, and soil conservationists say that some land now growing pasture and farm crops would be more suitable for growing timber. So, we are not running out of wood any time soon.

For another thing, trees are fairly efficient collectors of solar energy. Through the process of photosynthesis, trees convert sunlight into stored heat in the cellulose of the wood fibers—heat that is relatively easy to harvest

and store by those who use this energy source to warm their homes. For a good share of us, wood can be gathered from nearby, usually at reasonable cost.

For still a third reason, the hardware—fireplaces, stoves, insulated chimneys, etc.—needed to convert a dwelling to wood heat is readily available and easy to install.

Finally, and perhaps most important to those who can utilize wood for heat, this source of fuel gives us a personal energy independence, or at least reduces our dependence on outside sources.

In short, burning wood from our own trees, instead of burning oil from some Arabian shiekdom, is good for both the country and its citizens, so long as our wood producing "factory" is not abused by poor management.

Despite its potential as a plentiful, renewable source of heating energy, wood is not for everyone. The benefits of switching to wood accrue primarily to rural and small town Americans who live in the forested regions of the country. While wood may be a home heating bargain in rural Arkansas, it can be an expensive alternative for a Manhattan condominium dweller.

And this is the argument from which critics of wood get their best mileage. They look at where trees grow and where people live, and say that there isn't enough wood in the right places or at the right price for this energy source to ever become very important as an alternative to petroleum.

But the American Forest Institute points out that, at present, barely 13 percent of the nation's electrical energy is provided by nuclear power, even after a genera-

Few woodlots and forests produce wood fiber up to their potential; selective cutting and better management could triple wood production.

tion of technological developments and hundreds of billions of dollars in research. If we now supply more than two percent of our home heating needs from wood without really working at it, what might the country achieve if we put only a fraction of the research money into wood that has been spent on nuclear power?

We're learning more about using wood all the time. Better wood burning devices today are more efficient than the best built when wood was a commonly used fuel. Homes are better insulated and tighter against the wind and weather.

But the simplicity of choosing a wood burning heater that does the job for an individual situation is not what it was 40 years ago. In the 1930's, builders who wanted a fireplace laid up bricks or stone and mortar in a conventional, open throat fireplace that was little changed in design from those built hundreds of years earlier. When a man needed a new stove, the corner hardware store usually had two choices: a big cast iron, pot bellied model and a smaller, cheaper heater.

Today, more than 50 different types of wood burners are being built by some 800 manufacturers here in the U.S. and overseas. They are promoted—and priced—on the basis of certain features; often features that are mostly cosmetic. Some of the European and Scandinavian stoves, although good quality equipment, are overpriced compared with efficient models produced in the U.S.

Worse luck, there is not always an assured relationship between price and heating efficiency, regardless of the nationality of the equipment. Spending more money for a fireplace, stove or furnace does not automatically guaran-

tee that you are buying a more serviceable home heating device for your own situation.

You need not be overwhelmed by the variety of wood burners on the market, though. As you sift through the facts, figures and fiction, keep in mind this important question: Does this device meet your needs?

Will the wood burner be your only source of heat, or will it merely supplement your present heating system? How large and what shape area will you need to heat? Will you need a heating unit that can operate several hours without recharging? Can you incorporate a wood stove or furnace into your present forced air or hot water home heating system?

Figuring the kind of wood heating equipment you will need involves the same exercise we described earlier, to size the heating plant to your location, house size and layout. Many manufacturers of stoves and furnaces list the BTU output of their equipment. This heat output rating usually is based on burning wood of a certain quality, of course, as woods of different density and moisture content vary considerably in the amount of heat produced. This average BTU figure will give you a good starting point to work with when you are deciding what size unit to buy.

However, deciding which model of wood burner to buy is one of those exercises where the more information you gather, the more confused you can become. The problem soon becomes one of too much information and how to accurately sort through it.

Each manufacturer's literature tells you why his stove is the "best" or "most efficient", but doesn't always tell you much about the form and function that makes it the best stove for your purposes. By all means, read the available literature on different wood burners. Talk to dealers. Look at the different types of stoves and their dominant features. Visit friends and neighbors who burn wood. Ask them not only about the performance of their stove, but also about the dealer who sold it.

Chances are 50-50 that the dealer you choose will determine the brand and model of wood heater you eventually buy, rather than the other way around. A reliable, knowledgeable dealer can give you a lot of information not only on the equipment, but also on what type of chimney system and installation method will best suit your home and its heating needs.

But not everyone who sells wood stoves is capable of giving good advice, so you'll need enough knowledge of your own to evaluate what the dealer tells you. If a dealer cannot answer your questions clearly and specifically, and help solve your heating problems directly and honestly, reconsider doing business with him. A wood stove or furnace is too large an investment to risk making a wrong choice.

HOW WOOD BURNS

As wood burns, several things happen: (1) water is removed by evaporation; (2) the wood breaks down chemically into charcoal (carbon), gas and volatile liquids, with carbon dioxide and water vapor being the chief by-products; (3) some of the gases mix with incoming air and are burned; and (4) the charcoal burns, producing most of the heat.

One pound of dry wood (zero moisture content) of any species has a calorific value of about 8,600 BTUs. Any moisture in the wood reduces the usable heat by carrying heat up the flue as the water is vaporized, and each pound of water vaporized uses up about 1,200 BTUs.

Other available heat is lost in the formation of volatile liquids and gases during combustion, but this loss varies with the type of heating unit and is part of the efficiency factor of any stove or furnace.

Air dried wood typically contains about 20 percent moisture. Therefore, a pound of air dried wood contains 1/5 pound of water and 4/5 pound of solid wood, giving each pound the potential heat value of just over 7,000 BTUs. But remember: that's the *available* heat in a pound of wood. To get that much heating value from the wood, you'd have to burn it at 100 percent efficiency, and no equipment does that. How much of the available heat you recover depends on the design of the equipment you use and how it is operated. Of the many types and styles of fireplaces, stoves and furnaces on the market, efficiency can vary from a net heat *loss* to a high of 70 percent from a well designed unit.

Wood-burning device	Efficiency range
Open masonry fireplace	0 - 10%
Heat-saver type fireplace	10 - 30%
Box stove (single firebox)	20 - 35%
Airtight stove	40 - 60%
Highly efficient stoves and furnaces	55 - 70%

The BTUs you recover from a pound of air dried wood may range from 350, with an open fireplace, to 4,900 with an efficient airtight stove. That's quite a difference in heating efficiency.

The more efficient wood burners are designed to preheat the air going into the combustion chamber, or firebox. The burnable gases released by wood during the burning process must be supplied with ample oxygen at 1,000 degrees F. or more, with most of the air over and around the fuel, rather than drawn up through the fuel bed. The "Down-draft" type heating units are designed so that these burnable gases pass along a circuitous route—either around baffles or through a separate chamber above the firebox proper—where they mix with a current of heated air and nearly all burn. In less efficient devices, these gases escape unburned up the chimney.

Wood-burning devices come in all shapes, sizes and styles. This fireplace-stove can handle some cooking duties. (Mohawk Industries, Inc.)

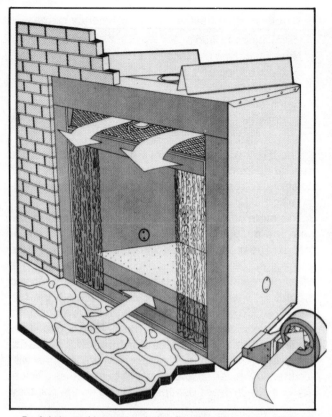

Prefabricated heat-circulating fireplaces are up to 30 times more efficient than open masonry fireplaces. Some, like this Western Fireplaces model, include warm-air circulating blowers. (A. R. Wood)

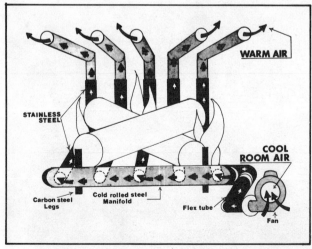

A combination grate and heat circulator can boost the efficiency of a fireplace considerably. (New England Fireplace Heaters, Inc.)

FIREPLACES

Few fireplaces—including Franklin type units—are designed with this efficient "after-burner" chamber. Most of them allow a large volume of heated air to escape up the chimney. An open fire is romantic and cheery and the stuff Christmas songs are made of, but it's not the way to go if efficient heat is what you're after.

In fact, several studies have proven that an open fireplace built into an exterior wall can actually remove more heat from a house than it adds. Here's the reason for that seeming paradox:

A roaring fire in an open fireplace radiates some heat into the room where it's located. But the oxygen supply to the fire is not controlled. The fire and the natural draft created by hot gases rising up the flue suck heated air out of the house, forming a negative pressure that pulls cold air into the house through cracks around windows, doors and other openings.

Installing a damper and glass fire screens with draft controls can increase a fireplace's efficiency, by controlling the supply of oxygen to the fire and by limiting the flow of hot gases up the flue. Special grates made of tubes that circulate warm air also help. Ducting combustion air from outside the house, to provide the fireplace with its own oxygen supply, can avoid having heated air from inside the house drafted up the chimney.

Perhaps the best solution for an old style masonry fireplace is to seal off the fireplace and install an efficient wood burning stove on the hearth. Several companies build small, compact units precisely for this purpose. In fact, USDA Secretary Bob Bergland, quoted earlier, adapted the fireplace in his home this way. Some of these stoves come with damper panel kits or backing plates to convert a fireplace flue to accept standard size stovepipe.

A freestanding fireplace does a somewhat better job of heating than does a masonry fireplace. Most of these— including the Franklin type—stand away from the wall, where air can circulate around the hot metal jacket and collect heat. This improves the *heating* efficiency of the unit somewhat, compared with a traditional open fireplace, but doesn't always do much for the *combustion* efficiency. Freestanding models with doors do allow some control of the flow of air to the fire, which results in better combustion efficiency.

Modern heat saving fireplaces, of the Heatilator® type, have metal side walls and backs with space for air to circulate between the walls and the combustion chamber. Air inlets near the floor and outlets at about mantel height provide convection air heating in addition to the heat radiated from the fireplace. Some units have small electric fans to circulate air through this heat jacket. These heat saver models also contain a tight closing damper and many of them have provision for using combustion

Free-standing fireplaces are better heaters than masonry units in an exterior wall; air can circulate on all sides of the metal jacket and pick up heat. (Goodwin of California)

air from outside the house. Some heat saver fireplaces are designed to be an insert form, around which brick, stone or other masonry is laid. In appearance, the completed unit resembles the traditional masonry fireplace; in operation, it is several times more efficient than an old open type fireplace.

More efficient fireplaces can be used to handle the heating demand in late fall and early spring, before winter weather really sets in, but most fireplaces are more ornamental than useful. If you're shopping for the most heat per pound of wood burned, don't even stop at the fireplace display. An airtight, draft controlled stove or furnace is what you're after.

Finding the Right Stove

If you're building a new home, or if your present home is equipped with forced air ducting or hot water heat piping, your best heating bet may be a wood furnace or combination fuel system. We'll talk more about these units shortly.

But if your goal is to find a stove that can be centrally installed to take care of all, or a major part of your home heating, look long and hard at the so-called "airtight" stoves, as most manufacturers of efficient wood burners describe their product. Actually, airtight is a misnomer, since fire must have some air to burn. These stoves are more correctly air *controlled*. That is, air enters the firebox only through a draft control and moves through designed passageways, rather than around loose fitting doors or through cracks in the stove's joints.

Because the amount and entry point of combustion air is controlled (either manually or by a thermostat) the rate of burning can be controlled accurately. As more air is admitted, the fire burns hotter. If the draft control is closed completely, the fire merely smolders, and eventually goes out.

Airtight stoves usually are constructed of either cast iron or steel plate. Cast iron is less likely to warp under high heat than steel, but is more brittle and subject to cracks under rapid temperature changes or impact. There's probably little difference between cast iron and steel, as far as heat holding ability is concerned—given an equal mass of metal. Stoves made of heavy steel plate and carefully welded should be as durable as the one built of cast iron. Some stoves are built with a steel firebox that is protected from warping by firebrick, refractory clay or steel liners. Most airtight stoves, whether constructed principally of cast iron or steel, do have cast iron doors. Even a slight amount of warping in a door could cause something less than an airtight fit.

Some airtight stoves have manual draft controls; others have bimetal spring thermostats that adjust the draft. In theory, the thermostat allows just enough combustion air into the firebox to keep the stove at an even preset temperature, and the better models do a good job. While a thermostatically controlled draft is somewhat more convenient, with experience you can learn to adjust a manual draft so that the wood burns evenly over a long period of time.

You'll read stove advertisements that talk about "secondary combustion" or "second stage combustion cham-

Round holes in this foundation are for ducts to carry combustion air to a corner fireplace from outside the house, rather than having the oxygen supplied by heated air inside the house.

bers" that are supposed to burn off the gases released by the wood, thereby increasing efficiency and eliminating creosote and tar problems in the chimney. Regardless of the claims, the truth is this: if combustion air mixes with these gases at 1,100 degrees F. or hotter, the gases burn. If the gases are not kept at 1,100 degrees or hotter until they are mixed with combustion air, they will exhaust unburned out the chimney, carrying with them tars, water vapor and part of the wood's potential heat. While the design of a stove has a great deal to do with its efficiency, the burning of combustible gases is governed by a law of physics that does not change with the brand of stove.

Most manufacturers of better-made stoves (Ashley, King, National Stove Works, New Hampshire Wood Stoves, to name a few) publish efficiency ratings on their equipment, but not all of them use the same benchmarks for computing combustion and heating efficiency.

The efficiency claims are somewhat like the Environmental Protection Agency's estimates of automobile miles per gallon: pretty good for comparing between different models, but not particularly helpful in telling you how well a unit will perform for you. Among airtight stoves of similar design and construction quality, efficiency will not vary a great deal. You'll also want to consider other features: safety, length of firebox (a shorter firebox requires shorter wood, which means more wood cutting labor), weight (a heavier stove holds heat longer), size of ashpit or drawer (a larger ashpit doesn't have to be cleaned as often), dealer reliability, appearance (if good looks is a factor) and price.

Three drawings showing different levels of efficiency in burning the wood gases in stoves which account for 40% of the heat potential in wood.

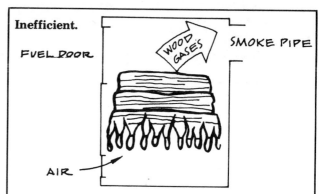

"Chunk" Stove—Air enters through ash door. Large volume of wood gases escape out smoke pipe.

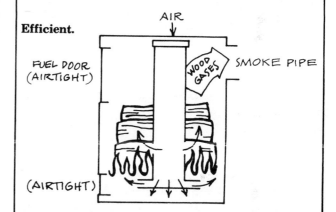

Controlled Draft—Air enters above, is preheated and spread across and above the fire. Some wood gases escape.

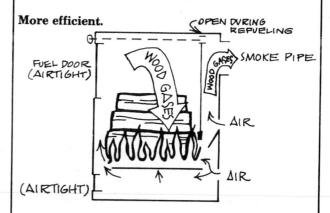

Controlled Draft and Sealed Fuel Chamber—Pull of draft and small self-generated pressure of wood gases in sealed fuel chamber force them into fire. Small amounts escape.

Circulating wood heaters have a metal jacket surrounding the firebox; air is heated in the space between and circulated into the room. These stoves are safer than other models to install in a room where young children play. (Ashley Wood Heaters)

A wood-burner can be installed in tandem with a conventional forced-air furnace system, to couple economy with convenience. (National Stove Works)

FURNACES AND COMBINATION SYSTEMS

In recent years, several manufacturers of wood burning devices have married the best of technology with economy to produce some remarkably efficient home and water heating implements. These devices use the automation and convenience of central forced air or hot water heating in tandem with the economy and versatility of burning more than one fuel—usually wood or coal, along with either gas or oil.

These wood burners can be roughly grouped into three main categories, although some of them fit into more than one slot:

• Central wood burning furnaces. These units are designed with a sheet metal plenum or jacket around and over the fire box and the same kind of ducting to different rooms of the house as with a gas or oil fired forced air furnace. An electric blower, controlled by a thermostat, pushes warm air through the system.

• Add-on or "piggy-back" systems. These are wood burning devices that can be channeled to the duct system of a conventionally fueled furnace; or equipped with internal piping for use with a radiator or hot water baseboard heating system, for hook up to a water heater for domestic hot water; or a combination of more than one of these functions.

• Multi-fueled furnaces. These central heating plants can use two or more dissimilar fuels in the same unit. These furnaces most often are built to use wood, plus either gas or oil, and may be equipped to heat domestic hot water as well as living spaces.

To describe each of these types of wood burners, we'll talk about the features of specific makes of equipment produced by well established manufacturers; not that these makes are better or more popular than others, but because the operation and application of them is fairly typical of all units of that type.

CENTRAL WOOD BURNING FURNACES

Ashley Wood Heater Company, probably better known for building automatic wood burning circulators, also makes an automatic wood burning furnace. This unit operates as any conventionally fueled forced air furnace, for up to 12 hours on a single charge of firewood.

The furnace has an airtight welded steel firebox, with cast iron liners and an insulated cabinet around the combustion chamber. Combustion air is controlled with a bimetal thermostat. A two-speed, 700 cubic-feet-per-minute (CFM) blower moves heated air through sheet metal ducting to registers in rooms to be warmed.

(The furnace can also be equipped with a smaller blower and used as a supplemental unit with gas or oil fired central heating systems.)

Here's how the furnace works: when the bimetal thermostat cools down, it opens a draft door to admit combustion air through a downdraft distributor to the firebox. The wood burns hotter and heats the air in the plenum mounted directly over the firebox. A pre-set temperature switch is mounted in the warm air plenum and wired to the electric motor of the blower. The automatic switch can be adjusted from 80 to 120 degrees F. When the heat from the firebox has warmed the air in the plenum to the pre-set temperature, the automatic switch turns on the blower to push warm air through the ducting system. As the space warms, the bimetal thermostat slowly reduces the amount of combustion air going to the firebox, the rate of combustion slows, the temperature in the warm air

A central wood-burning furnace functions just as any other central forced-air heating system—the major difference is the fuel used. (Ashley Heaters)

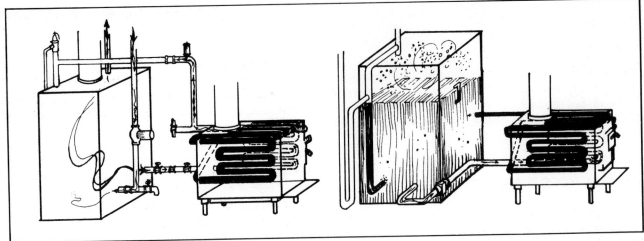

Typical "piggyback" installations with a wood stove's internal hot-water piping connected to a steam boiler (National Stove Works)

plenum drops below the pre-set point on the automatic switch, which shuts off the blower.

The combustion air damper can also be controlled by a remote thermostat; as in cases where the furnace is installed in an unheated basement, and the thermostat is mounted on a wall in the living room above.

The Ashley furnace takes firewood in 24 inch lengths, has an output of 75,000 BTUs when fired with seasoned hardwood, connects to a standard six inch flue opening and weighs something over 300 pounds. Several similar wood burning furnaces are on the market and—with some minor differences—operate essentially in the same way.

"PIGGYBACK" SYSTEMS

The Thermo-Control® line of wood burners built by National Stove Works, Inc., is made in three sizes that can be fitted with internal piping to augment a hot water (hydronic) heating system, to preheat water for domestic hot water, or both. The unit also can be equipped with a warm air plenum and ducted directly to a gas or oil fired warm air furnace, with the added option of interior piping for hot water. The accompanying sketches show typical hook ups for these add on or "piggyback" systems.

The largest model built by National Stove Works features a rolled steel firebox lined with firebrick, and takes

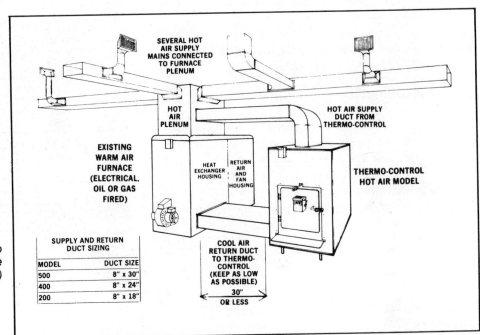

Wood stove connected to an existing warm-air furnace (National Stove Works)

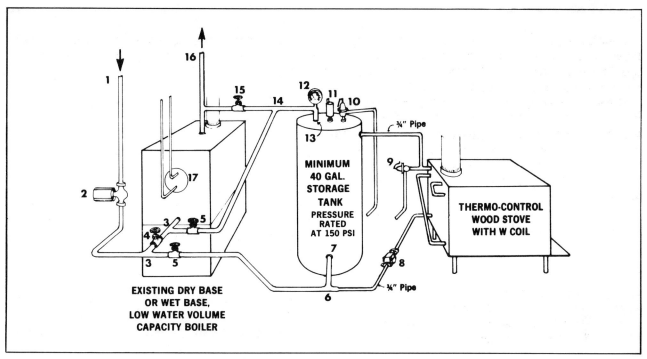

Example set-up with a wood stove interconnected with both a home heating system and a domestic hot water system. (National Stove Works)

wood in lengths to 28 inches. The combustion chamber (firebox), when full, holds about 270 pounds of hardwood fuel, and burns the wood at a rate of 5 to 18 pounds per hour, depending on the heating demand, to put out up to 135,000 BTUs per hour. The combustion air intake is controlled by a thermostat.

When these units are hooked up in tandem with hot water systems, boilers or forced air furnaces, the wood burner usually becomes the *primary* heating system and the conventionally fueled device is operated as a back up or auxiliary. The conventional system goes into service automatically when the wood burner is not operating or is not able to keep up with the load.

This have-your-cake-and-eat-it aspect of these piggyback systems is especially attractive for people who don't want to be tied to fire stoking and ash hauling chores of wood heat alone. During the week, you can operate the system on economical wood heat; then take off for a skiing weekend and let the conventional heating system take over, without worrying about your pipes freezing because the wood fire went out.

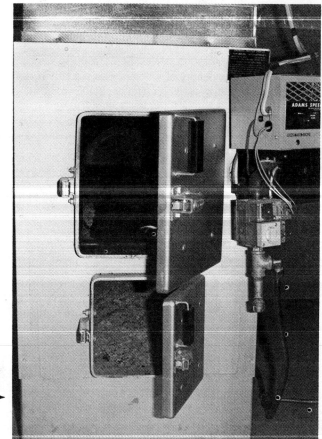

Dual-fuel Longwood furnace has a 5-foot-long firebox for ► *wood, and a 150,000 BTU gas burner just to the right of the wood loading door. The unit is completely automated—the gas takes over when wood is not burning.*

Multi-Fueled Furnaces

Dual fuel type furnaces offer a slightly different approach to the flexibility and economy of piggyback set ups. These are high output heating devices that can operate on wood and one or two other fuels.

The author's family home has been heated for the past three winters with a Longwood® combination furnace that uses both wood and L.P. gas (propane). The center of the furnace is a heavy steel 5 feet long firebox. At the front of the firebox, just beside the loading door, is a powered gas burner. A light gauge metal jacket sheathes the firebox and is connected to a warm air plenum above the combustion chamber. The plenum is ducted to living spaces and an electric blower pushes warm air throughout the house.

A centrally located thermostat is the master control for the entire system. Temperature limit switches control the main blower and a smaller blower on a heat reclaimer installed in the stovepipe between the furnace and the chimney. This heat reclaimer is ducted into the crawl space under kitchen and dining room floors.

In all but the coldest temperature, one filling of wood lasts 10 to 12 hours. When the firebox cools down, as when the wood has burned or at times when the door has been opened while fresh wood is added, the gas burner automatically comes on. If no wood is added, the propane tank assumes heating chores. If a fresh charge of wood is in the firebox, the gas burner only runs long enough to reheat the firebox and ignite the new wood.

If the dual fueled furnace were to operate *all the time* on gas, it would not function as efficiently as a furnace designed to be fired only with gas. But when operated as a wood furnace with a gas back up system, this type of heating device combines economy with convenience and automation.

Multi-fueled furnaces are also built with hydronic systems that heat living spaces with hot water radiators or baseboard convection heaters. Both hot air and hot water multi-fueled furnaces can be equipped with heating coils to preheat domestic hot water.

How much will it cost to install central wood heating? If your home already has a satisfactory chimney and warm air ducting in place, you probably can install a wood burning or multi-fuel furnace for something less than $1,000—no more than you'd expect to pay for a reliable gas or oil fired heating plant. Hot water systems may run more, even if you already have plumbing and radiators installed, simply because more safety and control equipment will be required.

Whether you're thinking of buying a Franklin type fireplace to warm up the family room, or a central wood burning system to heat the entire house and preheat the potable hot water, look around and compare different models available and the features they offer. Shop with an eye first for safety and efficiency; price second. Saving a hundred dollars by buying a cheap stove is not much of a bargain if the cut-rate contraption burns down your house.

YOUR WOOD SUPPLY

You've read this far on the subject of wood heat, so you probably have an accessible supply of wood that will make the use of this fuel practical, whether you buy firewood or cut it from your woodlot or other lands.

As noted earlier, the heating value of wood varies by the type of wood and depends largely on the density and moisture content. Any wood will burn, but denser (heavier) wood, if well dried, delivers more heat for a given volume of wood. And that's important to know, whether you are buying firewood or cutting your own. Wood is not often sold by weight, but by the *cord* (A cord is a unit of outside measurement of wood occupying 128 cubic feet of space, or a stack of wood four feet high, four feet wide and eight feet long).

On an air dried basis (20 percent moisture content) a cord of shagbark hickory wood weighs 4,160 pounds; while a cord of box elder weighs but 2,500 pounds. The potential heat (if you could burn the wood at 100 percent efficiency) in a cord of hickory is 29,100,000 BTUs; the available heat in a cord of box elder is 17,500,000 BTUs.

Buying Firewood

You can readily see that whether buying firewood or cutting it, the BTUs stack up a lot faster with hickory than with box elder.

It's a little like the gag question school kids ask: "Which is heavier, a ton of feathers or a ton of lead?" Of course, a ton is a ton—of anything. But the feathers make a much bigger pile. On a pound for pound basis, at the same moisture content, a medium density wood, such as box elder, contains about the same amount of available heat energy as does a denser wood, such as shagbark hickory, but it makes a bigger pile.

Let's suppose you plan to install an airtight wood burning stove with a rated heating efficiency of 60 percent, and hook it up to your oil fired furnace as an auxiliary natural heating system. An oil furnace typically operates at about 70 percent efficiency—seldom much better than that. If the fuel oil costs 80 cents per U.S. gallon, what can you afford to pay for air dried ash wood to get the same heating value per dollar?

APPROXIMATE WEIGHTS

Here are approximate weights of standard cords of various species of wood, when green (or freshly cut) as compared with the same wood air dried to 20 percent moisture content:

Species	Green (lbs.)	20% Moisture (lbs.)
Ash	3,940	3,370
Basswood	3,360	2,100
Box elder	3,500	2,500
Cottonwood	3,920	2,304
Elm, American	4,293	2,868
Elm, red	4,480	,056
Hackberry	4,000	3,080
Hickory, shagbark	4,980	4,160
Locust, black	4,640	4,010
Maple, soft	3,783	2,970
Maple, sugar	4,386	3,577
Oak, red	4,988	3,609
Oak, white	4,942	3,863
Pine, shortleaf	4,120	2,713
Red cedar	3,260	2,700
Sycamore	4,160	2,956
Walnut, black	4,640	3,120

Here's how to use this information when you're buying wood: Suppose you buy a cord of freshly cut ash wood. You're still buying 23,600,000 BTUs of available heat, just as if the wood were air dried to 20 percent moisture content. However, you're also buying 570 pounds of extra water that must be converted to vapor in your stove. Since it takes 1,200 BTUs of heat to vaporize each pound of water, a cord of green ash wood has about 684,000 fewer useable BTUs than a cord of air dried wood of the same specie. If you buy green wood, rather than air dried ash wood, you are losing the equivalent of five U.S. gallons of fuel oil or seven therms of natural gas for each cord—just in the energy needed to "boil off" the extra water.

You'll recall from Chapter 3 that heating devices ordinarily used with various fuels differ in their percentage of efficiency. But strictly on the basis of the *available* BTUs of heat energy, here are quantities of other fuels needed to equal a cord of air dried wood of various species:

AVAILABLE BTU's

Species	BTUs per cord (millions)	U.S. Gals. No. 2 oil	Therms (100 cu.ft.) Natural Gas	U.S. Gals. L.P.-Gas	Kilowatt/hrs. Electricity
Ash	23.6	168.6	236	259.3	6,941
Basswood	14.7	105.0	147	161.5	4,324
Box elder	17.5	125.0	175	192.3	5,147
Cottonwood	16.1	115.0	161	176.9	4,735
Elm, American	20.1	143.6	201	220.9	5,912
Elm, red	21.4	152.9	214	235.2	6,294
Hackberry	21.6	154.3	216	237.4	6,353
Hickory	29.1	207.9	291	319.8	8,559
Locust, black	28.1	200.7	281	308.8	8,265
Maple, soft	20.8	148.6	208	228.6	6,118
Maple, sugar	25.0	178.6	250	274.7	7,353
Oak, red	25.3	180.7	253	278.0	7,441
Oak, white	27.0	192.9	270	296.7	7,941
Pine, shortleaf	19.0	135.7	190	208.8	5,588
Red cedar	18.9	135.0	189	207.7	5,559
Sycamore	20.7	147.9	207	227.5	6,088
Walnut, black	21.8	155.7	218	239.6	6,412

At 60 percent efficiency, your stove will deliver only about 14.3 million of those potential 23.6 million BTUs in a cord of ash firewood. But that's the equivalent of the heat you'd get from 145 gallons of fuel oil, burned at 70 percent efficiency. At 80 cents per gallon, the oil costs $116.

Of course, that $116 pays for the oil delivered in a steady stream to your furnace burner—no wood to cut and tote, no ashes to haul, no dampers to adjust. And that kind of convenience is worth something; worth more to some people than to others. But purely on the basis of heat delivered—in this example, at least—a cord of air dried ash wood is worth up to $116, when fuel oil costs 80 cents per gallon.

On the other hand, a cord of cottonwood, with its 16.1 million available BTUs, would be worth only about $70. It would take 1½ cords of cottonwood to equal the heating value of one cord of ash wood.

Incidentally, in many parts of the country, huge commercial fuel yards are being developed, where wood is piled for sale to wood burning customers. Some of these yards sell wood by the ton, rather than by the cord, so the above tables can help you evaluate what you're paying for each unit of energy on that basis.

In other localities, wood sellers may price wood by the "rank" or "tier" or "face cord." These terms usually relate to a percentage of a cord, based on the length of the logs. For example, a "tier" or "rank" of wood cut into 16 inch lengths would be a stack four feet high, eight feet long, but only 16 inches wide—rather than the 4x4x8 foot stack that comprises a cord. In this instance, a "rank" of wood is a third of a cord. If the wood were cut into 24 inch lengths, the "rank" or "tier" would be a half cord. Unless you know the woodsman you buy from, it's a good idea to do your own measuring and convert the volume to cords. It keeps everybody honest.

Let's say, for example, that you buy a load of 24 inch long fireplace wood that forms a closely stacked pile that measures five feet high by 12 feet long. Multiply the height (5 ft.) times the length (12 ft.) times the length of the logs (2 ft.), and divide by 128. In this case, the pile of wood measures only 120 cubic feet, or just short of a standard cord. If the seller calls it a cord and charges you accordingly, you're being cheated out of eight cubic feet of wood. Some states have passed laws requiring wood sellers to provide a bill of sale stating the cords or fraction of a cord sold.

Be especially careful of buying wood in small lots, such as the "fireplace bundles" sold by supermarkets and other outlets. Seasoned oak wood weighs about 3,600 pounds per cord. So a bundle of oak wood that weighs 36 pounds is about 1/100th of a cord, and it's no heating bargain at $2.49.

Standard firewood measurements. A "face cord" or "rank" is a fraction of a cord, depending on the length of the logs.
(University of Missouri)

Where to Find "Free" Firewood

Depending on where you live, you may be fairly close to an abundant supply of wood that you can cut and haul for winter heat free of charge, or at a small cost.

For example, you can get a permit to cut up to 10 cords per season (for personal use) on U.S. Forest Service lands. In western states, the Bureau of Land Management often allows firewood cut on public lands. In addition, many state agencies allow limited wood cutting in state forests, state parks, game reservations and other public woodlands.

Contact the nearest Forest Service district ranger for a permit to cut wood in National Forests. The main reason for the permit is so the ranger can explain the type of wood to be cut—usually "down and dead" trees, or undesirable species that need thinning from stands of better timber. The same thing holds true of Bureau of Land Management lands.

For permission to cut wood on state owned land, check with your state forester, county extension agent or conser-

In many areas, firewood can be cut on public lands free or at a nominal charge.

Wood-cutting tools: Chainsaw, axe, sledge (some woodsmen prefer a splitting maul) and wedges.

vation agent. Some states have a free permit policy similar to that of the U.S. Forest Service, others charge a small fee. Most have restrictions on what, when and where you can cut—as well as how much.

Remember that cutting wood on public lands is a *privilege*, although your taxes help pay for the land and the salary of the guy who tells you when, where and how to cut the wood. Those lands are managed for all citizens, not just for the few who are able to cut firewood there. Abusing the privilege is the quickest way to lose it.

Often, you can find free or cheap wood by following timber crews. A sawmill logging crew is primarily interested in those logs that are large enough to be sawed into lumber. They leave a lot of firewood in the form of limbs, tops and defective trees that may be available for private wood cutters.

Telephone and power companies maintain crews to keep trees and limbs from growing into the wires. When these crews are working on rights-of-way in your area, you may be able to haul away the by-products of their maintenance efforts.

Lumber mills produce a great deal of scrap material—slabs, edging and trim—in milling operations. Often, the company offers these scraps at a small charge. While most slabs and scraps contain a high percentage of bark and sapwood that does not make premium grade firewood, these materials can be used along with better logs to stretch the supply.

Good manners and common sense are the most appreciated character traits when you're cutting wood on land that doesn't belong to you. Never cut wood without the permission of someone who is authorized to grant it. If you enter someone else's property without permission, it's called trespassing. If you cut wood on someone else's property without permission, it's called stealing.

Cutting and Curing Firewood

You need only a small investment in time and tools to cut all the wood you'll burn to heat your home. We will not go into wood cutting tools or how to use them, (*See Structures'* "Successful Homeowner's Tools") but a man with a chainsaw; an axe, a splitting maul and a couple wedges can spend a total of a week's time in the woods and easily harvest four or five cords of firewood.

Firewood should be seasoned (air dried) for at least six months from the time it is cut until it is burned. A year is better. We've already gone into some of the reasons for burning air dried wood, versus using firewood directly off the stump: you'll get a lot better performance from your heating equipment.

Of course, you can cut wood anytime. But wood cutting is the kind of activity that goes best with cooler weather, for both the cutter and the timber he cuts. Firewood cut in late fall, winter and early spring, when the trees are bare of leaves, will have slightly less moisture content than wood cut from trees that are busy pumping water to the leaves. Also, trees that are bare of leaves have less "sail" effect from any wind that may be blowing, and can be felled with more safety.

However, we are concerned here mainly with how you handle wood after it is cut. Properly stored firewood cures faster, stays drier—and burns better. The ideal storage is an airy, well ventilated, roofed woodshed large enough to hold a year's supply of firewood. But any method that allows wood to be stacked loosely, well off the ground, and covered to keep out most of the rain and snow will be far superior to merely dumping wood in a pile exposed to the weather.

For convenience, wood should be stored fairly close to the house; but not so close that termites, carpenter ants and other pests can easily migrate to the building's wood framing. Handling wood costs—if not in money, at least in time and effort—and the fewer times you have to handle each piece of wood, the better.

Some wood cutters like to saw wood into "pole" lengths in the woods, then haul these longer pieces to a central stack. Cutting and hauling goes faster this way. You have fewer cuts to make and fewer pieces to handle. Then, as time permits, the 6 or 8 foot poles can be bucked into lengths for burning. One objection to this system is that wood takes longer to cure. Wood dries partly by evaporating moisture from the ends of the logs; the more cut end surfaces exposed, the faster the wood dries.

Wood stacks can be covered with scrap lumber, old metal roofing or anything else that is cheap and will shed most of the rain and snow. Canvas or plastic covers can be used during periods of heavy rain or snowfall, but should be removed so the sun and air can dry the wood. These materials interfere with air circulation through the wood pile, unless some kind of framework is built to hold them up off the wood.

As a rule, it's not a good idea to store more than about a two day supply of firewood in the house. There's always the hazard that the wood may harbor rot, fungus or wood chewing insects that could infest the wooden parts of the building. An exception to this might be if wood is stored in a basement with concrete floor and walls, or when the weatherman predicts a blizzard.

You'll get more heat with fewer problems from firewood that is cut, stored and handled to provide a dry, accessible supply of fuel all winter long.

MANAGING YOUR WOODLOT

A man who owns a woodlot made up of mostly hardwoods that will supply heating fuel into perpetuity is a man increasingly to be envied. He has a measure of security denied those who are unable to wean themselves from the petroleum pipeline.

Woods land varies in its timber producing potential, with soil type, climatic conditions, growing season length and other factors. But, as a rule of thumb, an acre of "wild" hardwood trees should produce a sustained harvest of one-half cord per acre, per year. Woodlots managed more intensively may supply a cord or more of wood per acre each year, on a permanent, sustained basis. So, if you expect to burn five cords of wood each heating season, a 5 to 10 acre tract of hardwoods should provide your heating fuel from now on.

Managing a timbered patch of land for fuelwood production is about as much art as skill, for most woodlot owners. But it isn't that difficult to learn if you pay attention to the telltale signs Mother Nature reveals about the condition of individual trees. Be selective in your harvest. Cut first those trees that, when removed, will give more room and nutrients for the growth of more desirable trees in the stand. Cutting only the straightest, fastest growing trees (they are easier to split and build a woodpile quicker) and leaving the unhealthy, crooked, limby trees can soon reduce any woodlot to a "junklot."

You need not be completely ruthless about it, of course. It's neighborly to leave standing an over-the-hill hollow tree as a den for wildlife, and it's eye pleasing to spare a gnarled and twisted dogwood for the promise of blossoms one more spring.

But, generally, you'll need to manage the woodlot not just for the firewood it yields this year and next, but for the fuel you'll harvest 10 or 15 years from now. For instance, would you spare a badly formed, stunted black walnut tree and cut a straight grained, healthy white oak growing nearby? The oak will make the best firewood. And the walnut, should it ever grow into a prime veneer log, could be worth several dollars per board foot. But in this case, the twisted, runty walnut tree will never grow into anything of much value, and may even die prematurely. On the other hand, the white oak promises to continue to grow rapidly and increase your bank account of heating fuel each year. Cut the walnut and leave the oak.

Here are some things to keep in mind as you set priorities on cutting trees for firewood:

1. Harvest sound dead trees right away, before decay processes reduce the amount of useful wood in the tree. Utilize limbs and tops from logging operations, wind damage, etc.
2. Harvest diseased or insect infested trees, and burn the brush and remove the wood from the woodlot as soon as possible to prevent the pest from spreading to other trees.
3. Remove brushy, crooked or broken hardwoods.
4. Cut out "wolf" trees; those with unusually large, spread-

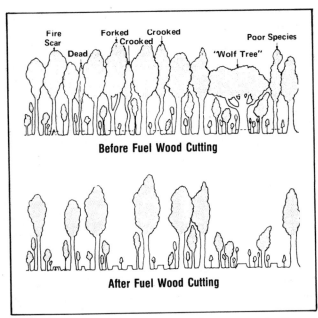

Fire Scar — Dead — Forked Crooked — Crooked — Poor Species — "Wolf Tree"

Before Fuel Wood Cutting

After Fuel Wood Cutting

Selective cutting can make any woodlot more productive. (University of Missouri)

Trees cut in late fall or winter, after leaves have fallen, provide wood that is slightly lower in moisture content than wood cut in spring or summer. (Note the blaze mark that indicates the tree was selected earlier for cutting)

ing tops that occupy more space than they deserve.

5. Cut "weed" trees of undesirable species.

One common error of woodlot owners is removing the "understory" of small trees and sprouts so the area looks neat and clean. Forget about what neighbors and passers-by may think of your woodlot housekeeping. That understory is an important part of your stand, and will grow to produce the firewood you'll need to heat your home 10 or 20 years from now. Meanwhile, those small trees and shrubs shade the soil and prevent it from drying out and blowing or washing away.

Safety First

A mass of wood burning in an enclosed firebox can generate temperatures of several hundred degrees. This much heat and burning material *can* be a potential fire hazard.

However, wood heat is relatively safe, if some precautions are taken to install equipment properly and operate it safely. Virtually all home fires that result from wood heating are caused by unsafe installation of either the heating device or its chimney, or by the careless operation of the equipment. Very few are caused by defects or faults in the stove or furnace.

Before you install wood burning equipment, check with your local building inspector and notify your insurance agent. Your home insurance premium may increase somewhat. But if you use wood burners *without* notifying the insurance company, and a fire—of whatever origin—does occur, settling the claim may be quite a hassle.

Installing the equipment properly in the first place is the major contribution you can make to safety. Wood burning devices can be used safely with only two types of chimneys: the familiar brick or stone masonry chimney, or prefabricated insulated metal chimneys (approved for use with "solid fuels" or "Class A" fuels) such as those built by Dura-Vent and Metalbestos. Thin wall light gauge stovepipe is not meant to be used as a chimney, but merely to connect the stove or furnace to an approved chimney.

If the stovepipe must pass through a wall of combustible material to connect the stove to the chimney, a ventilated "thimble" at least three times the diameter of the pipe should be used. Never run single wall stovepipe through a ceiling or into a concealed space, such as a closet or attic.

Wood burning equipment should be installed with a minimum of 36 inches clearance from combustible surfaces, such as wall paneling. If the wood burner must be installed closer than that to a combustible wall, a heat shield should be used between the stove and

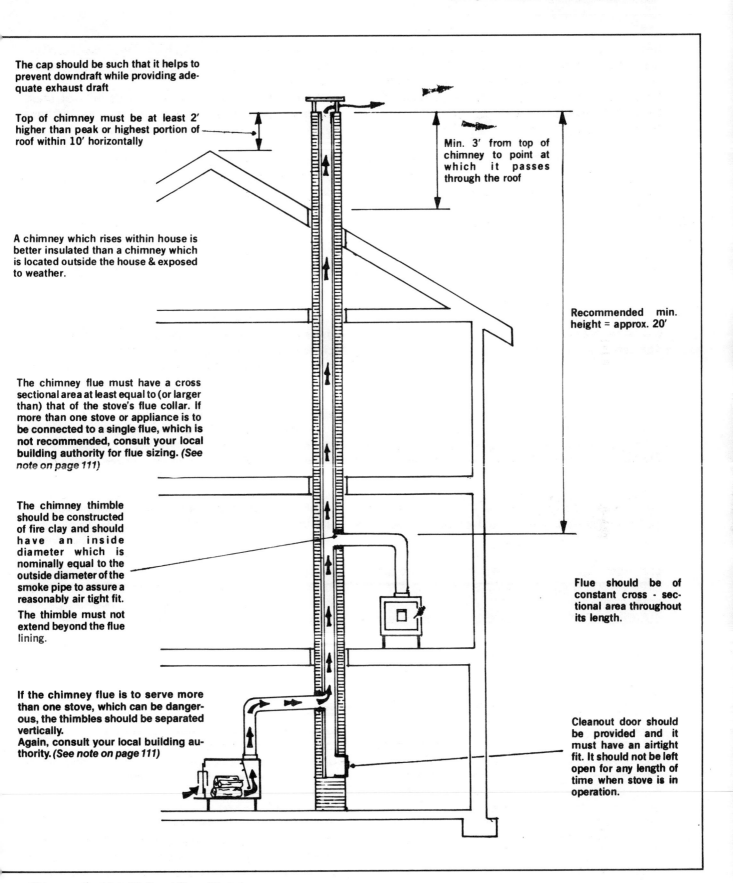

The cap should be such that it helps to prevent downdraft while providing adequate exhaust draft

Top of chimney must be at least 2' higher than peak or highest portion of roof within 10' horizontally

Min. 3' from top of chimney to point at which it passes through the roof

A chimney which rises within house is better insulated than a chimney which is located outside the house & exposed to weather.

Recommended min. height = approx. 20'

The chimney flue must have a cross sectional area at least equal to (or larger than) that of the stove's flue collar. If more than one stove or appliance is to be connected to a single flue, which is not recommended, consult your local building authority for flue sizing. (See note on page 111)

The chimney thimble should be constructed of fire clay and should have an inside diameter which is nominally equal to the outside diameter of the smoke pipe to assure a reasonably air tight fit.

The thimble must not extend beyond the flue lining.

Flue should be of constant cross - sectional area throughout its length.

If the chimney flue is to serve more than one stove, which can be dangerous, the thimbles should be separated vertically.
Again, consult your local building authority. (See note on page 111)

Cleanout door should be provided and it must have an airtight fit. It should not be left open for any length of time when stove is in operation.

Chimney checklist. (National Stove Works)

the wall. Some circulating stoves and stand alone fireplaces can be safely installed closer to a burnable surface, but make sure that claim is verified by an independent, recognized testing agency (such as Underwriters Laboratories, which affix the "UL" seal) before you risk it.

Combustible floors also need protection from the heat of a stove or furnace. Ready made "stove mats" of sheet metal and insulating material serve this purpose well. Sand, masonry or metal can also be used. The floor protecting material should extend 18 inches past the sides and back of the stove, and 24 inches past the front (or side) where the door is located.

Once it's installed safely, take your time getting acquainted with your wood burner. Don't build roaring fires right away. Build a small, easily controlled fire and—while it is burning—check stovepipe and chimney connections. Experiment with the draft control dampers to learn how the fire behaves at different control settings. Always open dampers or other combustion air controls before starting a fire in a stove. Use paper and dry kindling to get the fire going; NEVER use lighter fluid, kerosene or other flammable liquids in a wood stove. NEVER burn artificial logs of compressed sawdust or other materials in a stove—these are meant to be burned in a fireplace.

Earlier, we mentioned that gases and tars released when wood burns can condense inside a chimney to form deposits. "Airtight" stoves and furnaces with restricted air supply and the resulting low rate of combustion can cause more of these materials to form in chimneys than would be deposited by stoves and fireplaces with a faster rate of combustion.

When soot, tar and creosote build up in a chimney, they often form a hard, crystalline layer of almost pure carbon. If the coating ignites, the result is a chimney fire—the dread of all wood burning folk.

A chimney fire is a spectacular sight. As the carbon burns, temperatures inside the chimney may go as high as 3,000 degrees F., far beyond what any chimney is designed to withstand. The intense fire sucks oxygen into the chimney with a roaring rush (the roaring usually is the first warning that a chimney fire is in progress) and spews flame and sparks out the top of the chimney like some giant blow torch.

A fire inside a chimney can cause a lot of damage to the flue; can set the whole house ablaze, for that matter. Prevention is the best cure. Once the deposits inside a chimney ignite, there's little that can be done to control the fire, other than try to contain it to the chimney. If (heaven forbid) your family ever experiences a chimney fire, get everyone out of the house—fast—and call the fire department. Close dampers and draft controls on the stove to reduce the supply of oxygen to the fire in the chimney. Then stand by outside with a bucket of water or garden

hose and watch for fires that may start on the roof of the house, as sparks and burning chunks of soot spew out of the chimney.

Fortunately, chimney fires are usually of short duration—five minutes or less. A chimney fire will have burned itself out by the time the firemen arrive. If it hasn't set the house afire, the firemen will usually inspect the chimney, roof and attic of your house for any lurking danger of fire, and for damage caused by the chimney blaze. At any rate, don't use the wood burner again until the chimney has been thoroughly inspected.

Obviously, it's the better side of wisdom to avoid a chimney fire; and you can, with a little preventive maintenance that starts with the way you operate your fireplace, stove or furnace. Use seasoned dry hardwood and operate the wood burning device as the manufacturer recommends.

You can develop some operating habits that will reduce the deposit of tars and soot inside the flue. When you recharge the firebox with a fresh supply of wood, the firebox and chimney cool down considerably. For one thing, opening the door lets in a slug of cooler air. For another, you have loaded the stove with a mass of wood that is well below combustion temperature.

Open the draft for 15 or 20 minutes after reloading the stove, to let more air into the firebox and get the new wood burning. Allowing a greater volume of air into the firebox also causes smoke and gases to be expelled up the chimney faster and in greater volume, which carries much of the offending tars and water vapor on out to the atmosphere. Then, when the fresh wood is well caught, re-set the draft control to its normal operating position.

Perhaps the worst fire management technique is to toss green or wet wood on top of low coals or a low burning fire. Some people do this, in the belief that the wet wood will dry out and ignite. And it will, eventually, if the fire doesn't go out first. But before it burns, the wet wood will give off water vapor and condensable materials. As the fire is not burning hot enough to create a strong draft to carry these materials out the chimney, they build up inside the flue.

With well designed equipment connected to a properly installed chimney and fired with the right kind of wood, a one-a-year mechanical cleaning should keep the chimney in good shape indefinitely. Most of the material that adheres to the flue lining will be soft, crumbly soot and fly ash which can be brushed loose fairly easily.

But if you're just getting acquainted with a new wood burner, climb up and inspect the chimney fairly often for awhile. Note the kind and extent of deposits that build up in the flue with different fire rates and different kinds of wood. If gummy or hard material starts to build up, it usually collects first at the top of the chimney or where the chimney exits the roof, where the flue is cooler.

Flue cleaning is a messy job at best. One way to dislodge soot and other materials is to run a length of medium-sized chain up and down in the flue occasionally. (Photo by Jim Anderson)

Thermostatic heat detectors, installed in the room where the stove is located, sound an alarm when heat rises to about 135 degrees F. For more protection, also install smoke detectors elsewhere in the house. (Courtesy of Westinghouse)

If this material builds up to the point that a quarter inch or more of gummy, tar-like substance coats the inside of the chimney, it should be cleaned out. To date, the only way to do this is to brush, chip and scrape at the material, and that's quite a job. But it's several times less of a problem than a chimney fire.

Several chemical chimney cleaners are on the market, such as Red Devil® and Beacon® brand products. These cleaners are used in wood burning equipment while a moderate fire is going, and when carried into the flue with the smoke and gases, they react chemically to cause soot and tar deposits to flake and break loose from the flue lining. At least, that is the claim for these products. Experts disagree on how effective chemical chimney cleaners are, but most agree that they should be used only with a relatively clean chimney.

Other safety precautions with wood burning equipment fall mainly under the categories of good housekeeping and common sense:

• Don't place combustible objects near wood-burners.

• Don't let coals and sparks fall onto rugs or wood floors while loading in wood or cleaning out ashes.

• Use metal containers to remove ashes, and take the ashes—hot ashes—out of the house. Ashes can insulate coals and embers that will continue to glow for a considerable time.

• Don't burn large amounts of paper or cardboard in the stove or fireplace. Some people open Christmas presents in front of the fireplace, and feed in the wrappings. A few even stuff the old Christmas tree into the fireplace. And a good number of those people no longer have the fireplace, or the house it stood in.

It's a good idea to locate a fire extinguisher handy to each wood burner. A pressurized, dry chemical type extinguisher costs little, and can be used on all classes of fire.

Other safety devices you may want to install—whether you heat with wood or not—are heat and smoke detectors. These detectors sense either the smoke in the air or an increased temperature and sound a high pitched horn or beeper alarm. They come in both plug in and battery powered models. Heat detectors utilize a pre-set thermostat (usually set to activate the alarm at about 135 degrees F.), while most smoke detectors use either photo-electric devices or radioactive ionization detectors. As most home fires produce considerable amounts of smoke (most residential fire deaths are due to smoke inhalation), a smoke detector should be one of the warning devices installed. Some homeowners who heat with wood install a heat detector in the room where the stove or furnace is located, and install one or more smoke detectors in other parts of the house.

A properly installed fireplace, stove or furnace, operated with care and common sense, can be a safe, economical source of home heat. If you have any questions about installing or using a wood burner, don't hesitate to get professional help and advice. Wood is an old time heating fuel, but a generation or more of Americans have grown up with heat provided by gas, oil or electricity. Don't feel embarrassed if you don't know all there is to know about burning wood.

There is a paragraph in Chapter 10 of *Successful Fireplaces*, Structures Publishing Company, which, in the interest of safety, must be quoted: "Old timers will tell you about running several stoves into one flue. Don't. With modern low-cost factory-built chimneys, there is little reason why you should take this risk. National Fire Protection Association, 470 Atlantic Ave., Boston, MA 02210, in their booklet, recommends against it. But should you wish to live dangerously, the NFPA provides a table of square inches of flue capacity related to combined BTU input of the stoves and height of the flue."

Most identifiable source of geothermal energy for many Americans is Old Faithful geyser, in Yellowstone National Park. (Courtesy of U.S. Park Service)

7
THE ENERGY IN MOTHER EARTH

"In the eleven western states, more than thirty million people could benefit from the development of geothermal space heating."

Paul J. Lienau
Geo-Heat Utilization Center

Since the earth was born as a white-hot coal billions of years ago, only a comparatively thin crust of the planet has cooled. The amount of heat energy stored in the molten rock that lies beneath the earth's surface is virtually limitless—and so far, practically untapped.

Thermal energy stored in hot rocks, hot water and underground steam amounts to a lot more than merely a happy accident of geography for a few people. The most readily available underground heat is in areas where the earth's mantle is thin, as in regions with active volcanoes and earthquakes. But new drilling and heat transfer technology is pumping heat from deeper and deeper within the earth, and moving it greater distances on the surface for space heating, domestic hot water and industrial processing.

"Over ten percent of the U.S. population resides within 40 miles of known geothermal resources," says Paul J. Lienau, of the Geo-Heat Utilization Center, Oregon Institute of Technology. "Here at Klamath Falls, wells 300 feet deep average 200 degrees F."

And the resource is not limited to the western United States. There are at least 1,000 heat-flow drillings on the east coast, and several known geo-pressure areas along the Gulf of Mexico. Juvenile geologic regions in Hawaii and Alaska promise still more energy. The U.S. has more than a thousand known hot springs that potentially could be developed.

In fact, geothermal energy is a fast developing source of heat world wide. About half of the homes in Iceland are heated by geothermal water, some of it piped from 900 feet below the surface. Australia, New Zealand, Italy, Russia, and Japan all are known to have major geothermal sources.

Unfortunately, knowing that energy exists below the earth's crust in almost unlimited quantities is not the same thing as harnessing that energy source. For practical purposes, the heat stored in Mother Earth is used in one of two ways:

1. Underground steam is piped to generating turbines to produce electrical power.

2. Underground hot water and steam for direct heating and industrial processing is piped to the surface, or, heat exchangers are lowered into "hot" wells to extract underground heat.

Both of these techniques are usually expensive, involving much the same kind of technology as developing an oil well, and requiring more dollars than most homeowners can afford. Geothermal energy has a great deal of potential as district-wide resources, however, where underground heat is used to generate electricity or provide space heating for whole neighborhoods or communities.

In non-geothermal areas, the moderating effect of the earth itself can be put to more immediate use by homeowners as in earth contact or underground buildings; or with heat and "cool" extracted from the earth or underground water. We'll go into more detail on these energy uses of the earth shortly.

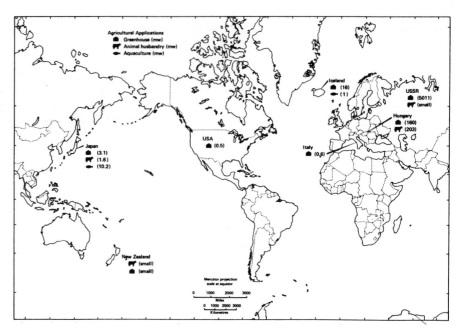

Geothermal regions of the world show developments to use Mother Nature's underground heat for various agricultural and industrial applications. In addition, geothermal energy is considered a medical resource by some people.

GEOTHERMAL ELECTRICITY

There are vast quantities of underground steam and pressurized hot water in the earth, with considerable potential for generating electricity. The known and estimated geothermal energy in the U.S. could be tapped to produce from 7,500 to 15,000 million watts by 1990 (about two to four percent of the estimated total demand by that time).

To date, however, most electricity generated by geothermal energy is produced by underground "dry" steam, and there aren't a great many of those fields known at present. Most geothermal wells yield hot water, or a mixture of water and steam. The cost and efficiency of generating plants, designed to use this "wet" steam, are not nearly as attractive as those which can operate on steam at high enough temperature and pressure to vaporize all the water.

The largest geothermal electric plant in the world is in the U.S. It is in an area in northern California known as "The Geysers." Here, about 100 miles from San Francisco, the Pacific Gas & Electric Company operates plants that produce about 700 million watts of electricity. The Geysers have enough potential energy to produce at least triple the present electrical output.

Even with "dry" steam, geothermal generation of electricity is not particularly efficient, unless some use can be made of the waste heat released by the process. But then, that is true of plants powered by fossil fuels and nuclear energy.

While plants such as those at The Geysers can deliver centrally generated electrical power, at costs that are competitive with other steam powered generating installations, it's not a source of electricity many homeowners can develop on their own.

GEOTHERMAL SPACE HEATING

Theoretically, geothermal energy could be reached at any temperature up to 6,000 degrees F., by drilling deeply enough. Practically, the hot interior mass is too deep (in most parts of the world) to reach with existing drilling technology.

Still, the potential for geothermal energy broadens considerably when you consider the direct use of this energy source for space heating, domestic hot water, and many industrial processes. As much as 30 percent of the U.S. space heating could be supplied by the relatively abundant 130 to 200 degree water within drilling range of the earth's surface.

In addition, thermally "marginal" water in the 60 to 100 degree range could be used in water-to-air or water-to-water heat pump systems. These systems could be installed about anywhere in the U.S., with wells developed either on a district-wide or individual basis, where ground water at 60 degrees or above can be pumped to the surface. The principle of using a heat pump in conjunction with water, that is warmer than the outside air temperature, is similar to that discussed in Chapter 4, where we describe combining heat pumps and solar systems.

"Space heating at temperatures below 212 degrees F. is by far the largest single energy use at temperatures suitable for direct geothermal applications," says Lienau.

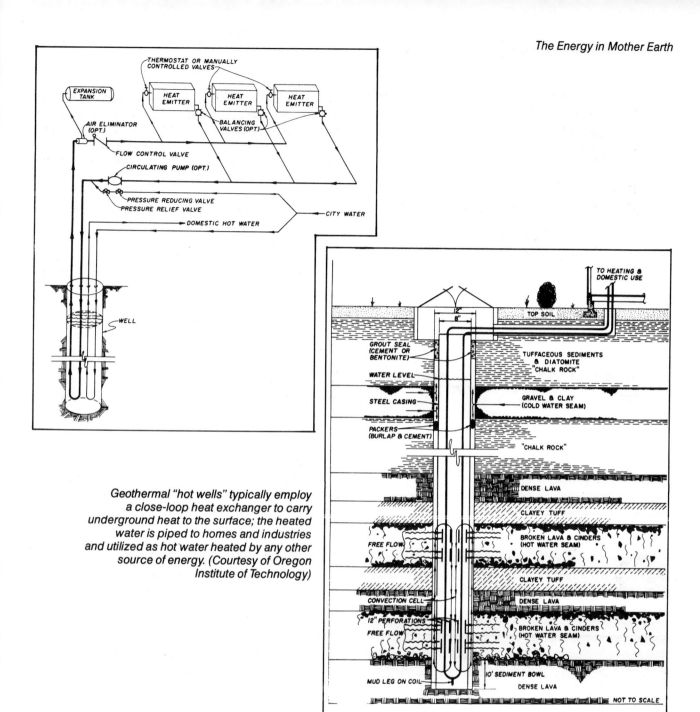

Geothermal "hot wells" typically employ a close-loop heat exchanger to carry underground heat to the surface; the heated water is piped to homes and industries and utilized as hot water heated by any other source of energy. (Courtesy of Oregon Institute of Technology)

"We have the technology now to utilize geothermal energy for these direct applications. But, more advances are needed to reduce the costs and increase the distance over which geothermal energy can be competitive with other sources of heat."

In Klamath Falls, Oregon, more than 400 wells are used for space heating, using water from 105 to 200 degrees F. The principal heat extraction system is a closed loop down-the-well heat exchanger, using city water in the loop. The entire campus of Oregon Institute of Technology is heated by geothermal water at an annual operating cost of about $30,000. That's a savings of nearly $250,000 per year, compared with the cost of heating with conventional fuels.

Around the world, in areas where geothermal sources are near the earth's surface, individual homeowners are developing systems to heat living spaces and domestic water. And, where water at the right temperature can be found close to the surface, this is an economical alternative to using conventional heating. But right now, the main emphasis on direct geothermal heating is on more efficient, large scale district heating projects.

In the national energy picture, the use of direct geothermal energy in food and industrial processing has the

Earth-berming and below-grade construction require special design expertise. Building in contact with the earth saves heating energy in winter; cooling energy in summer. (National Solar Heating and Cooling Center)

potential to save a great deal of conventional fuel—and to save a great deal of money for corporations able to use this source of energy.

BELOW-GRADE HOUSING

Not all of us live within range of geysers and hot wells. But there are more practical ways to make use of that most plentiful of all natural resources: the good earth itself.

Civilized man grew up in subterranean shelters—caves, dugouts and other below-grade living quarters—that were warmed in winter and cooled in summer by nature. And he's going back—at least some people are. More and more home builders are snuggling their dwellings into the protective bosom of Mother Earth.

Earth-contact housing is being evolved in several different building styles, including: earth bermed (soil piled up on the sides); earth covered (where even the roof is covered with three feet or more of soil); and inset into hillsides (with excavations into steep grades and earth back-filled against walls).

Now, don't think of earth-contact housing in terms of dank, dark basements you may have known. Construction standards and methods—by which comfortable, healthful below-grade living quarters are being built—will erase that image of underground living quickly. (We'll discuss some of these features more thoroughly in Chapter 10, where the Davis and Mason families describe living in underground and earth-contact homes).

While there are some definite benefits, there are also some design challenges to a comfortable, livable earth tempered dwelling. Because of the obvious need for stronger construction to support the pressure and weight of earth in contact with walls and roofs, earth-contact housing can cost more—but probably little more, if any, than well-built conventional housing.

For most, it's a radical departure from conventional housing, and some families who have built below-grade homes have encountered special problems along the way. Our purpose in this section is to help would-be earth-contact builders side step some errors before the house is constructed. Once a structure is anchored with tons of soil, it's tough to correct any flaws in design or construction.

First off, how much can you save on energy by moving below grade? That depends on a good many things, and no one is waiting in the wings to make money back guarantees. However, Mary Tucker, housing specialist at Kansas State University, estimates *at least* a 10 percent savings in energy, compared with well built above ground housing.

The earth is more of a temperature "shock absorber" than an insulator. The massiveness of earth slows down temperature changes, so that there are daily and seasonal lags. In one study, the lowest average outside air temperature for the site was in February, while the lowest average temperature nine feet below the ground surface was in late May. The warmest outside air temperature was in August, while the warmest temperature at nine feet underground occurred in November.

In other words, summer heat doesn't get through to the lower levels of a below-grade house until November—just when you're beginning to need it. And the cold arrives at that level in May, just when outside air temperatures are beginning to warm up. That's why underground water pipes sometimes freeze in late spring, after the soil on the surface has thawed and begun to warm up.

Of course, the thermal conductivity of a particular soil may vary. Wet, sandy soil conducts heat 10 times as fast as dry sand. But, all soil has the ability to absorb and store a considerable amount of heat as it attempts to pass through. This characteristic, plus the isolation of walls from the chilling effect of winds, accounts for the major energy saving benefits of below grade housing.

Further, grass and other flora growing on the soil acts as a barrier to heat flow, by creating air pockets, reflecting solar radiation, and transpiring moisture from the soil.

"One of the more glaring mistakes being made by families building earth-contact homes is that they pay too little attention to the soil type for their areas," says Ms. Tucker. "The soil type is important also because it helps dictate the type of drainage system you need. If the drain system is not designed for a specific site and soil type, you can get into problems with water inside the house."

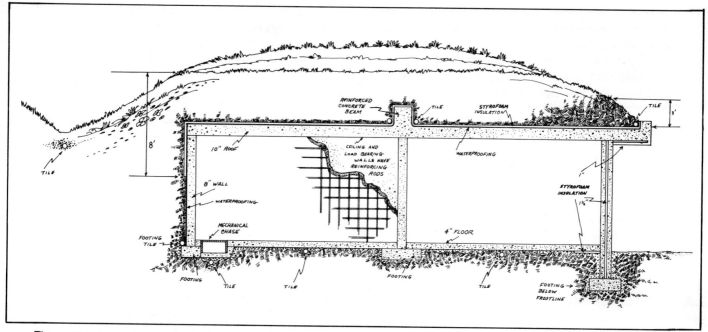

The earth acts as a temperature "shock absorber" in an underground house. (Courtesy of Davis Caves, Inc.)

The Walls

Lateral pressure is a major consideration when designing earth embanked walls, and this pressure depends on fill depth and properties of the soil. More lateral load means increased wall cost, in most cases, and any below-grade design must balance the trade offs between depth of fill, insulation value, and cost of building a stronger wall.

Earth fills of 8 to 10 feet, common to residential structures with basements, can usually be retained without much extra cost. But deeper fills may require more cost, depending on the height of the wall. Some of the backfill pressure and "overturning" forces can be countered by designing geometric shapes to create a mass that resists these forces: round, hexagonal, trapezoidal, etc.

"If you're building concrete walls, pay close attention to the quality of the concrete used," says Ms. Tucker. "Before pouring, insist on a quality material worked as a stiff mix and vibrated in the forms to prevent honeycombing. Excess water in concrete can cause void spaces that invite leaks."

Some earth-contact builders use lightweight, pressure-treated, wood foundations that can be built on the ground as a "tilt-up" wall, or prefabricated in panels elsewhere and transported to the building site for assembly. This can be cheaper than concrete, but it's not a construction method that tolerates shortcuts.

You'll want to make sure that the wood is treated with the proper chemicals to resist decay and insect damage. Suppliers of CCA treated wood products often warranty the material for an extended period of time. A crew of three or four men is needed to handle 4 by 10 foot wall panels. For larger wall sections, a crane or other mechanical lift may be needed.

Wood has good beam strength; lightweight wood walls can resist soil pressure well. But they must be held in place by well designed floor and roof structures which serve as horizontal beams. Also, the exterior surface of a wood wall placed in contact with earth must be carefully waterproofed.

Ventilation and Moisture Control

The ventilation system chosen should be determined by the type of earth-contact house built, and probably should be designed to provide several complete air changes daily. However, the lowest *adequate* ventilation rate is probably the best, particularly when dehumidifiers are used to control condensation within the structure.

Warm dry, air although it has low *relative* humidity, may still have a high *absolute* moisture content. Even on hot, dry days, outside air entering the building may cool below its dew point (condensing temperature) when it comes in contact with cool surfaces inside the house.

Moisture from condensation can be a hazard, especially in warm seasons when below-grade spaces remain at low temperatures, relative to outside air. Each change of air means an increase in house moisture and thus an increased load on the dehumidifying or air conditioning system. Ample insulation on exterior walls helps keep the temperature of below-grade walls above the dew point.

As mentioned earlier, proper design, good location and a good drainage system are necessary to keep ground and surface water out of below-grade dwellings.

Grading is among the first, and perhaps a more important step in moisture control. The finished grade should be sloped greater than one percent away from the house on all sides for a distance of at least 12 feet. For conventional roofs over earth bermed structures, gutters, flashing and downspouts can collect water from the roof and direct it away from the building. Roof overhangs of two to four feet help keep moisture from the walls, particularly from those walls facing prevailing winds.

For earth covered roofs, water falling directly on top cannot always be diverted as handily. These roofs need the extra protection of a continuous vapor proof membrane to shed water from the roof into a drainage system, or hold it in the soil until the moisture is evaporated or used by growing plants.

With all below-grade structures, a porous fill—such as stone or gravel—is needed behind walls to intercept ground or surface water, and a perimeter drainage system should be used to carry this water into a sewer or other removal system. The perimeter drain should be 6 to 12 inches below the footing to insure a positive flow of moisture out of the surrounding soil. In some cases—and with some soils—drainage lines also may be needed under the interior of the structure to carry off water that passes under the footing.

Walls below grade need to be waterproofed, both inside and out. In above ground structures, it's common practice to put a vapor barrier only on the inside surface of the wall, to stop moisture from migrating from the warmer interior into the cooler insulation in the wall. But with earth-contact buildings, when the soil is warmer than the interior of the house, moisture can flow *into* the structure. That's why vapor barriers are needed on both interior and exterior sides of walls (and ceilings, in underground houses) to stop moisture migration from both directions.

Insulation

Experts disagree on the amount of insulation needed in an earth-contact structure; but most agree that some insulation is required.

Wood is a reasonably good insulator, and below-grade walls built of treated wood have some resistance to heat flow. Also, the stud framing allows space for insulation material to be installed within the wall.

Masonry walls are most often insulated with rigid foam panels bonded to the exterior surface. Be sure that you use a material that is resistant to moisture. This allows the massive wall to serve as thermal storage. Some builders also install furring strips and conventional wall coverings over the interior wall surface, and add insulation to the space between the furring.

Heating Below Grade

The amount of heat needed by an underground or earth-contact house depends on the climatic conditions in the area, the soil type, and how much of the house is underground. Generally, the heating requirement of a below-grade house will be 10 to 30 percent less than the

Careful waterproofing is vital if treated wood walls are to be placed in contact with earth. (Courtesy of Bill Mason)

energy needed to heat an equivalent amount of above ground space.

The big heat conserving features of below-grade housing are: living space that's protected from wintry winds (less infiltration of cold air) and a mass of earth that absorbs much of the shock of rapid temperature changes. If the outside temperature is 10 degrees F., and you maintain the interior of a conventional house at 70 degrees F., that's a heating difference of 60 degrees. If the average temperature of the earth fill outside a below-grade house is 45 degrees F. when the outside air temperature is 10 degrees, you have a heating difference of only 25 degrees to maintain inside temperatures at 70 degrees. Burying all or part of the house in the earth has the effect of reducing the *heating degree days.*

Many below-grade builders incorporate solar heating with conventional heat systems as back-ups. A common design for earth-contact houses built into south slopes includes a solar collector on the exposed south wall. In these homes, the wall exposure is often a function of collector size.

Others install wood burning heating equipment. Don Hess burns about two cords of wood per winter to heat his 2,000 square foot underground home in Johnson County, Kansas. Hess notes that when he was building the home, before any supplemental heat was used, the temperature inside never got below 60 degrees F. even in January.

Some owners of below-grade homes build in heat "scavenging" systems to capture and store excess heat from appliances, kitchens and bathrooms—heat that normally is exhausted to the atmosphere. The soil around a below-grade structure forms a "heat sink" that can store a great deal of the heat that would otherwise be wasted.

In these systems, the rock fill behind the walls—needed primarily for water drainage—serves as an air duct through which exhaust air passes before it is released to the atmosphere. Heated air gives up part of its stored energy to the rock, increasing the temperature of both the rock and the earth that surrounds it. This slows the heat loss through earth embanked walls.

Dirt on Top, Too?

It's a good idea to get reliable structural design help before you build a structure to withstand the pressure of tons of dirt piled over and against it. Obviously, it's no job for amateur builders—and not one for just *any* concrete contractor—to build a ceiling to support a three foot deep covering of earth that weighs 375 pounds per square foot.

Several systems are employed to obtain the structural strength needed in totally underground homes, but the surest method is to pour concrete walls and roof in one monolithic structure. Wall and roof forms are bolted to-

gether in place; steel reinforcing is carefully located; and the whole job is completed in one continuous pour.

Discussing the engineering needed to design for the stresses encountered in building totally underground structures is beyond our scope here. It's a critical consideration that demands qualified counsel.

WOULD YOU LIKE LIVING UNDERGROUND?

Interestingly, most families who live in well designed, well lighted below-grade housing say it is a comfortable, enjoyable experience that spares them a lot of the noise and vibration to which an above ground house would be subject.

Some builders of below-grade houses report trouble lining up lenders to finance construction. On the other hand, they usually find that insurance companies are willing to write coverage at more favorable premium rates. A home built mainly of concrete and protected by several feet of soil poses fewer fire hazards. Also, earth covered structure is a natural storm shelter, and that's especially desirable in areas where high winds and tornadoes are hazards.

Underground Cooling

In most of the U.S., temperatures range a year-round 55 to 65 degrees F. at 10 feet or so below the ground surface. Innovative homeowners are beginning to explore systems to take that coolness from the earth and move it into the home to cut air conditioning costs.

Others use constantly cool underground water to cool living spaces. For instance, the Robert Redford ranch manager's solar home, described briefly in Chapter 4, pipes cold spring water through a coil inside the heating-cooling air duct. Air blown across this cooling coil gives up some of its heat and thus cools the living space.

Bill Enter, of Mustang, Oklahoma, mounts an oscillating lawn sprinkler on the ridge of his roof in summer. A solenoid valve, controlled by a timer, allows the sprinkler to spray a fine mist of well water on the roof at pre-set time intervals.

"We try to operate the sprinkler just enough to keep the roof shingles moist," says Enter. "The evaporating water pulls heat out of the attic. Our central air conditioner runs about half as much as it does when we don't use the sprinkler."

In another earth-as-cooling system, loops of 6 inch plastic pipe are buried 6 to 8 feet below ground. Warm air from inside the house is circulated through this loop,

A small lawn sprinkler operated on a timer-controlled solenoid valve sprays a mist of water on this roof, allowing evaporative cooling to pull heat from the attic, resulting in 50 percent lower air-conditioning bills.

pushing earth cooled air into the building's interior. As the earth adjacent to the pipe absorbs the heat from inside the house, this system loses its effectiveness. But with a long enough loop of pipe buried deeply enough in the earth, this method could provide much if not all the summer cooling load in many areas.

A somewhat farther out twist on this same idea, for summer cooling, is a man made "ice cave," similar to the ice storage houses used in New England and the Midwest years ago. An insulated cellar is filled with drums or other containers filled with water. Cold winter air is circulated through the structure to freeze the water. (It's vital to use containers that will not burst as the water freezes and expands). In this system, a fan is positioned to blow cold outside air through the "cave" in winter, and the cold is preserved in the structure until it is needed in summer for space cooling.

If the cold storage structure has a sufficient mass of ice, and is insulated well enough so that the ice does not melt before the air conditioning season comes, the system can pull a lot of heat from warm air inside the house as it is circulated through the cave.

Jerrold Parker, an electrical engineer at Oklahoma State University, has designed "earth storage" systems to be used in both heating and cooling seasons. Drilled deep wells—200 feet deep or more—are filled with water that is heated in winter by solar collectors. The warm water is circulated through heat pump heat exchangers. The heat pumps operate much more efficiently when drawing heat from the warm water than they would if operated to extract heat from cold outside air.

In summer, the storage wells are filled with cold water, and the heat pump—operated in the cooling mode—works with this cooler medium to cool living spaces more efficiently.

I repeat, not all of us live within range of a geyser or hot water wells, but all of us live close to the weather moderating influence of Mother Earth. Whether you're building *on* the good earth, or *in* it, there are a number of ways to make use of this universal energy-saving resource.

8
SHOULD YOU PRODUCE YOUR OWN "JUICE"?

"In areas served by a utility company, you may be able to sell your surplus power to the utility, which makes your investment in small scale power generating equipment even better."

Bill Kitching
Small Hydroelectric Systems
Arlington, Wash.

America runs on electrical energy. It powers our homes, schools, offices, factories and public buildings. As critical as gasoline is to the on-the-go American, motor fuel ranks a distant second to electrical power in energy importance.

If you want to learn just how dependent modern man is on electricity, switch off the current at the main entrance panel in your home and see what a short time passes before someone has a critical need for an implement that runs on watts. It's mind boggling that such a silent, invisible, odorless source of power could be so universally important to modern living.

With the climbing cost of all forms of energy—and the very real possibility that the supply of energy to some areas may be limited before long—more homeowners are looking into ways to generate at least part of the electricity they need from natural, renewable, close-at-hand sources.

And the interest is not limited to rural Americans. Wind powered generators are in use in the heart of New York City, and such other urban areas as Evanston, Illinois. In lock step with this interest, the market has become dizzy with hardware that lets you grab watts from the wind, water, directly from the sun, and from other renewable sources.

For most of us, the more promising near-term alternatives involve the conversion of the mechanical power in wind or water to electricity. However, direct conversion of solar energy to electricity (through the use of "solar cells") and the use of external combustion, low heat machinery

to generate power are steadily coming into the range of practical use, with new developments and lower cost manufacturing techniques. The technology, both home-made and commercial, runs the gamut from strictly "junk-yard" variety set ups to well engineered, sophisticated power generating machines. (The reference to "junk-yard" technology is in no way derogatory—it's amazing what a confirmed tinkerer can do with a homemade wind turbine and the generator from a 1949 Studebaker. The knowledge and experience gained from trial and error home built power systems can be invaluable. If the Wright Brothers had stuck with proven technology, they'd still be building bicycles.)

Actually, the technical know-how to build durable, serviceable small scale automated power generating plants has been around for a long time. Small hydroelectric systems that were installed more than a half-century ago still crank out electricity. Before the advent of rural electric cooperatives, many farmsteads were powered by low voltage Jacobs, Zenith and Wincharger units, or Delco plants driven by small gasoline engines.

Back then, power demands were small compared with nowadays. A few light bulbs and a radio; perhaps a small refrigerator and a water pump—these were the extent of the electrical demand in most rural homes a couple of generations ago. And going back to more limited electrical consumption may be part of the answer for any homeowner who installs his own generating plant today.

Installing an individual generating plant was relatively expensive back in the Thirties—even in terms of 1930 dollars—and the cost of producing your own "juice" has at least kept pace with inflation since then. But we have reliable control equipment today that frees a householder from some of the chores of watching gauges and needles that were part of his grandfather's price for energy independence. In particular, the solid-state electronic technology that fell out of the space program is being put to good use in small scale generating setups.

Today, there are probably as many different possible power generating setups as there are people interested in building or installing them. The automotive and camping industries have given birth to a variety of 12 volt DC (direct current) appliances, from lighting to air conditioning to television sets. Aircraft manufacturers build many appliances in the low to medium voltage range. And some homeowners are going to small output, low voltage systems, using the alternate source of power and storage batteries to operate outdoor lighting, essential or emergency circuits in the home, and other equipment. A few are going entirely to low voltage DC systems, and in some remote sites, where the cost of bringing in commercial power would be prohibitive, a low voltage generator-battery system may be the most economical way to go.

But for our purposes, we'll stick to plants smaller than 10 kw capacity that produce 110-120 volt, 60 cycle alternating current (AC—where the electrical current changes direction in the wire 60 times each second) which has become the standard for commercial power generating stations. In many cases, this involves a small generating plant that would supplement off-the-pole power by providing some part of an average household's electrical demand.

There are two basic ways of doing this: with a battery storage system and power converters, or with induction generators or synchronous inverters that allow the private system to be interconnected with the utility power grid.

WHAT'LL IT COST?

"Even with off-the-pole power costing up to 10 cents per kilowatt-hour, most private generating systems are hard to justify strictly on the basis of economics," says William Hughes, electrical engineer, Oklahoma State University. "At best, you're talking about $6,000 to $8,000 per 1,000 watts of capacity for a dependable system."

However, as commercial electricity rates approach or pass 10 cents per kw-hr (kilowatt-hour), these plants look more and more attractive. Also, an energy conscious government will help pick up the tab for some systems, in the form of income tax credits and sales tax rebates. Here's the way the arithmetic might go:

Let's say that a Vermont householder installs a 2,000 watts wind powered plant that will supply most of the electricity he needs. If we use the lower figure in Hughes' estimate of initial cost, the set up would involve about $12,000, for wind turbine, generator, tower, batteries, control equipment, etc.

Amortizing the equipment investment over 10 years, the capital investment breaks down to $1,200 per year. If the family uses an average of 720 kw-hr per month (and that's a fairly realistic average, with no electric heat or air conditioning) the power costs about 14 cents per kw-hr. That's considerably higher than commercial electricity current costs in most areas.

Also, if the homeowner borrows the $12,000—or part of it—there are interest payments to make. And it probably is not realistic to assume that any mechanical or electrical equipment will operate trouble free for 10 years. So, if we add another two cents per kw-hr generated, to cover interest, operating and maintenance costs, the rate comes to about 16 cents per unit. That's twice the cost of "boughten" electricity. . . or more.

However, the picture is not necessarily that gloomy. For one thing, the federal government allows up to a $2,200 deduction, as income tax credits, on the purchase of wind generating equipment. In addition, some states allow credits on sales tax or state income taxes for energy equipment purchases and installation. The Vermonter in our example can deduct $1,000 from state taxes as energy credits. That's a total of $3,200 that can be subtracted from the initial $12,000 investment, since these credits can be deducted directly from income taxes due. This brings the net capital cost of the system down to $8,800. Based on a 10 year life for the equipment, that comes down to $880 per year, and reduces the cost of power generated to about 12½ cents per kw-hr, when two cents are included for interest, maintenance, and operating costs.

Assuming that the homeowner pays for the system outright, there are other benefits to be figured. If the homeowner purchased the 720 kw-hr per month (8,640 kw-hr per year) from a commercial utility at six cents per unit, he would need to first *earn* $518.40 to pay the bill. That $518.40 income is subject to state and federal income taxes. If he's in the 25 percent tax bracket, that means he must earn $691.20—before taxes—to have the $518.40 available with which to buy power. Of course, to be realistic, the homeowner should figure the "opportunity cost" of the $12,000 he invested in a power plant. If he had invested the money in stocks, bonds, or passbook savings, the interest income might amount to more than $691.20 per year.

But it's a pretty safe bet that energy costs will increase faster than savings and loan interest rates. If commercial power goes to 15 cents per kw-hr in five or 10 years, the

Vermonter with his windmill is dollars ahead of the game.

Further, the National Energy Act of November, 1978, requires that public utilities, in some cases, must buy "surplus" power generated by private installations. In some hook ups, the utility allows the private generator to feed power into the power grid, through the electric meter, which causes the meter to run backward—thus crediting the account by the amount of commercial power generated by the private wind plant. In other instances, the utility enters into an agreement to purchase excess power generated at some wholesale rate, and requires a separate meter be installed to measure the power supplied to the utility grid.

Before you attach any local power generating equipment to a utility's meter or service line, get the company's permission. The company probably will require that positive disconnects are installed, to protect linemen working on company lines and equipment.

"We're grateful for any sources of additional power," says Sam Houston, general manager, COMO Rural Electric Cooperative, in central Missouri. "But we are concerned about the safety of our people who are working on lines and equipment. We want to know that no one is feeding into a line we think is dead."

We'll have more on the tax and money side of energy in Chapter 11. Suffice it to say that, with present utility rates in most regions, about the best an individual can hope for in the way of return on investment, in a generating plant installed where commercial power is available, is to break even. But money alone is not the only motivation for installing private power generation equipment. There's also a strong "what if" factor in most cases:

• What if the calamity that befell the Three Mile Island nuclear plant in Pennsylvania becomes an epidemic that causes us to lose the 10 to 12 percent of our electrical power generated by this source?
• What if fossil fueled generating plants can no longer get enough gas, oil or coal to operate?
• What if the cost of electricity goes up faster than the current projections of five to 7.5 percent per year?

If any or all of these possibilities come to pass—or come to pass sooner than expected—those homeowners who have invested in private generating plants, to provide at least part of the electricity needed, will be ahead of the game. They will have: an operating installation, a good deal of experience under their belts, and will be in much better shape to weather power "brownouts" or "blackouts" than will those who depend entirely on off-the-pole power.

CO-GENERATION SYSTEMS

In the past few years, equipment has been introduced to make home generated electricity compatible with commercial alternating current. Called "synchronous inverters", these solid state electronic devices use incoming commercial voltage and frequency as a reference for

Entertech's 1500 wind generator is compatible with utility power. (Courtesy of Entertech)

controlling the locally powered generator's output. In other words, the inverter matches the output of the generator to alternating current at standard line voltages and frequencies, to operate in parallel with commercial power (often called power "co-generation").

These systems do away with the need for storage batteries, DC to AC inverters, and other control equipment. When the generator is putting out more power than the electrical load requires, the excess flows into the utility company's power grid. If the generator is producing less power than required, the balance is supplied by the utility in the normal through the meter fashion.

Synchronous inverters built to handle normal household loads do not come cheap. Prices typically start at $2,000 and go up. But for private generating plants that will be interconnected with commercial utility power, they do away with the need for much equipment normally associated with a home power plant.

These inverters do not make wind or water powered generators operate indefinitely without attention or maintenance, but they free the homeowner from the frequent attention to the plant and its performance, and eliminate the need for battery storage and maintenance.

However, since the synchronous inverter is tied directly to the power grid, the output of the generator cannot be used when the utility is out of operation. This is an automatic, built-in safety feature, for both the generator and for servicemen who may be working on utility lines, but it prevents the use of the private generator during power outages. The power plant could be designed with battery storage and inverters for use in emergencies, but this would boost the cost of the total system out of practical reach for most homeowners.

Some companies are beginning to build generating systems designed specifically for homeowners who already have utility service. These systems, such as the Entertech 1500 wind generator, are meant to supplement commercial power, rather than provide the bulk of the electricity needed. The Entertech 1500, for example, is a wind powered induction generator that produces 115 volts of 60 cycle AC power at all operating wind speeds.

The induction generator idea is not new. This type of generator operates on the principle that an electric motor, when driven faster than its normal motorized speed, starts to generate current. For years, equipment such as elevators and oil well pumpers have used electric motors that re-generate as the equipment over drives the motor on the "coasting" stroke.

However, modern induction generators are much more efficient than electric motors that are mechanically driven past their operating r.p.m. Units such as the Entertech 1500 can be plugged into a 20 amp household wall outlet, where the power generated flows directly to lights and household appliances on the downstream side of the utility meter. When the wind doesn't blow or the water doesn't flow enough to carry the load, the utility automatically picks up the slack. These systems also incorporate an automatic shutdown feature to prevent back feeding into a dead utility line. The hardware for generating half, or more, of the electricity needed by an average household is reduced to the generator (plus tower, with wind plants) and the control panel.

The Entertech 1500, which produces up to 1.5 kw, will cost $5,000 to $7,000 installed, depending on the height of the tower.

BATTERY STORAGE SYSTEMS

For a dwelling already served by an electric utility, where the generating plant will be equipped with a synchronous inverter or other equipment to "tie" with the utility grid, forecasting the electrical demand is fairly simple—and only necessary from the standpoint of figuring out how much of the demand can be met by the generator. But where a battery storage system is planned, forecasting electrical demand may be more difficult—certainly it's more critical—if most of the power will be inverted from DC to AC.

These systems are, basically, battery charging plants. The generator charges a bank of storage cells with DC power through a voltage regulator. The DC power then is changed to 115 volt, 60 cycle AC current through an inverter.

The amount of electricity needed per month and in peak loads must be accurately estimated to design the size of the generator, battery storage, and inverter. A typical U.S. household consumes between 8,000 and 12,000 kw-hr of electrical energy per year. You may want to refer to the table in Chapter 3, for average power consumption of various lights and appliances. But for your own calculations before installing a battery storage, you'll need to get more specific information about how *your* family uses electricity: when, how much, and in what pattern.

Batteries are expensive. So are inverters, the equipment needed to impose a sine wave on direct current power to convert it to 60 cycle AC. For example, a 110 volt, 360 amp/hour battery will cost in the neighborhood of $3,000. Inverters cost about $2 per watt capacity and require nearly a third of their capacity just to run themselves. That's why sizing a system and all its components is important with a battery storage system.

The best way—the only *sure* way—to calculate your electrical requirements is to multiply watts times period of use for each light, motor and appliance. It's tedious and time consuming, but if you plan to provide all or most of your electricity from a battery equipped system, it's a necessary exercise.

Electrical usage must be accurately estimated with a battery storage-inverter system.

Every light bulb, every electrical appliance and motor has a power rating—usually indicated in the watts required to power it. A 100 watt bulb requires 100 watts. If you burn the bulb for three hours each night, you will require 300 watt hours per day, or nine kilowatt hours (9,000 watt hours) per month. Some motors may show the rating in amps on the nameplate data, rather than in watts. Don't let this throw you. Watts equals amps times volts. A 115 volt appliance with a 10 amp rating requires 115 times 10, or 1150 watts. This same formula—Watts X Amps X Volts—can be used to find any one of the missing numbers. For example, a 75 watt bulb requires .65 amps of 115 volt power, but requires 6.25 amps in a 12 volt system.

You'll also need to compute the *peak* demand for AC power. In a system with only lights, and appliances without motors, the demand is the sum of the power ratings of all electrical devices operated at the same time. Also, in those systems where a separate DC circuit is taken directly from the batteries to power DC only or "universal" AC-DC devices, peak demand is not so critical. The bat-teries can deliver virtually all of the power in storage more or less instantly.

But, where the size of the DC to AC inverter is a limiting factor, the capacity of the inverter must be adequate to meet the "starting" demand by AC motors on refrigerators, washing machines, furnace blowers, etc. Start up requirements for motors (called the "locked-rotor amperage") may be three to six times greater than their normal running requirements—that is usually shown in the nameplate data.

If you are planning to install a battery equipped generating system, give a lot of thought to using as many DC lights and appliances (directly from the batteries) as is possible and convenient, even if this means wiring some separate circuits for DC equipment.

In some generating systems, the energy is stored as heat or compressed air, as well as in batteries—or in lieu of batteries.

Finally, look into the practice of your local utility of setting favored electric rates for "off peak" use. More and more companies are considering higher rates for peak use periods—as during mid-afternoons in air conditioning season—and lower rates for off peak times. This may involve juggling the times you normally perform some household chores that consume electricity. But, if you can operate a home generating plant to go "off line" during peak use times, you may be able to negotiate a lower kw-hr rate for the commercial power you buy during non-peak times.

Selecting Battery Capacity

A storage battery is required for generating plants where energy must be available to the load on demand, but the power source is intermittent or insufficient to meet demands. Generally, lead acid batteries are most economical; but automobile batteries are not suited to the charge/discharge patterns of a power plant. A battery with a low self-discharge, such as a pure lead cell or lead calcium grid, is preferable.

Batteries are built with a value of 2.4 volts per cell. Thus, a 12 volt system would require six cells; a 24 volt system would require 12 cells; and so on. Sizing batteries at a rounded off two volts per cell allows for a charge/discharge efficiency of 90 percent.

Batteries are sized by "ampere hour" rating, whatever the combined voltage of the cells. For example, a 50 amp/hour battery can discharge one amp per hour for 50 hours, and—theoretically, at least—can discharge 50 amps for one hour. While the voltage of a battery is determined by the number of cells, the amp/hour rating depends on the number of plates per cell and the size of the plates themselves.

Batteries should be stored where the temperature can be kept above freezing. There are two reasons for this: (1) Batteries deliver more of their stored power quicker when the electrolyte fluid in them is warm, and (2) if batteries are discharged below about 50 percent, there's always a danger that the electrolyte will freeze and destroy the battery.

Also, from a safety standpoint, batteries should be stored where there are no sparks or open flames in the same compartment. Batteries under charge can give off explosive gases.

POWER "BLOWIN' IN THE WIND"

Wind is only solar power once removed. Air is set in motion by the uneven heating of the earth's surface, and carries a great deal of force.

There's a lot of potential energy even in slow moving winds. Efficient electricity generation depends not so much on high winds as on *steady* winds. In fact, the operating range for most wind powered generating plants is from about 7 to 30 m.p.h.—well below gale force breezes.

Obviously, the place to start planning a wind energy generating system is by measuring the resource, and we'll get to that shortly. But first, it's theoretically possible to extract only about 60 percent of the potential energy in the wind at best, and perhaps half of this is lost to slippage, gearing, drive mechanisms and design faults. Still, that amounts to a lot of work from a resource that is free and ever renewable. And it means that a great deal of potential electricity is riding those breezes that blow only 10 to 12 miles per hour.

Wind propellers of whatever type function by geometric law. If a 20 foot diameter turbine produces two kw in a 10 m.p.h. wind, a 40 foot blade of the same design in the same wind should produce eight kw—a fourfold increase, rather than merely twice as much. Also, the power available in the wind is the cube of the velocity. For instance, a propeller that delivers 500 watts in a 10 m.p.h. wind will produce about 4,000 watts in a 20 m.p.h. wind. So, you can see that doubling the diameter of the propeller (also called "rotor" or "turbine") results in the production of nearly four times as much power. But, doubling the wind speed results in an *eightfold* increase in power.

Of course, materials and design limitations of wind generating machines will not let that kind of geometry project indefinitely. At some point, the blades of the propeller start interfering with each other; the turbulence created by one blade slows the blade behind. The most efficient ratios are when the propeller tip speed (or the peripheral speed of the turbine) is 5 to 10 times that of the wind speed.

Most machines have a designed "cut in" wind speed (at which they start generating electricity) of 6 to 10 m.p.h. To protect the equipment in high winds, feathering and braking devices are designed to limit the speed of the rotor.

Since the performance of a wind generator is so affected by wind speed (up to the designed maximum), it's important to locate the wind plant where it will take best advantage of prevailing winds. In most areas of the country, the wind blows from the same, or nearly the same, direction 65 to 70 percent of the time. These are *prevailing* winds, and should average at least 10 m.p.h. for good power generation. However, the other 30 to 35 percent of the time, storm winds gust from other than prevailing directions. These winds, although they blow only two days out of seven, on the average, contain about two-thirds of the power in wind. Usually, rotor diameter is sized to produce the maximum output in these more powerful, though less frequent winds.

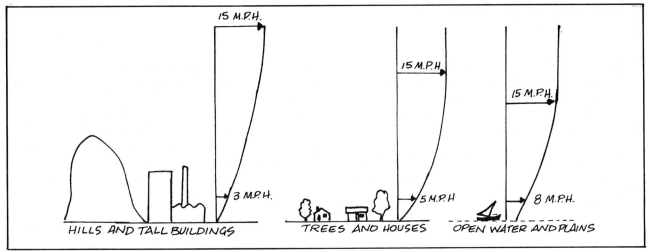

Winds are caused primarily by uneven heating of the earth's surface, but local topography can affect breezes in any particular area.

Evaluating the Site

Do you have a good site, with enough wind to operate a wind generating system? As noted above, an area with an average wind speed of 10 m.p.h. is a good candidate. But even that statement assumes that the wind will blow more or less steadily most of the time—say from 6 to 15 miles per hour. An area that is calm for six days, then has a 70 m.p.h. gale on the seventh day is not ideal, even though the wind speed may average 10 m.p.h.

The first step in selecting a good wind power site is to eliminate the clearly unsuitable ones. Trees, buildings and other obstacles cause wind turbulence; as do land contours—such as cliffs and steep, rough hills.

More promising sites are near flat, level ground and open water, where the wind is unimpeded by physical obstacles. Also, some topographical features can increase wind speed. A valley that stretches in the direction of prevailing winds between two hills can "funnel" the wind and increase its speed in a sort of venturi effect. Wind also speeds up where it crests a ridge or flows over a smooth hill.

Because of friction with the earth, wind speed is reduced near the ground surface; and increases with height, often in an almost linear curve. For example, if the wind speed at eye level of a person standing on the ground is 10 m.p.h., the speed at 40 feet high will be about 13½ m.p.h. and at 90 feet above the ground, will be about 15 m.p.h. Remember; the same wind plant will produce almost four times as much power in a 15 m.p.h. wind as at 10 m.p.h.

The Weather Service of the U.S. Department of Commerce hourly records wind direction and speed at several stations around the U.S. Local airports and weather bureaus also can be helpful in providing general wind speed and direction. But the only way to evaluate any potential wind plant site is to measure the wind at or very close to the wind rotor site. In other words, if you plan to install a wind plant in your back yard on a 40 foot tower, measure the wind there, rather than on or near the ground under the planned tower site.

A handy way to measure and compare wind speeds is to use an anemometer and odometer positioned at the site. These devices cost about $150 to purchase, but some wind plant manufacturers rent them to potential customers.

The odometer records one count for each mile of wind run through the anemometer's cups. Simply divide the number of counts per hour by 60 to find the average wind speed in miles per hour.

Naturally, the longer period of time you measure wind speed at the site, the more accurate planning information you have. It would not be realistic to measure the rainfall in April, then multiply by 12 to find the yearly precipitation. By the same token, it's not very accurate to measure the wind speed for only a week or a month, then multiply by 52 or 12 to find out how much power producing wind you can expect in a year's time.

However, if you have a weather station—or stations—fairly nearby, you can measure winds at your site for a long enough period of time (at least a month) to establish a correlation between your winds and those recorded by the weather station for the same time, then interpolate for the rest of the year. This is not as accurate as a full year's readings at the site, but takes much less time and work than measuring the wind hourly for a full year. And, generally, this will be a close enough estimate for planning purposes.

The local weather station will also have information on maximum winds and gusts that may be useful as you plan the kind and size of wind plant for your site.

Take the historical wind data provided by the weather station, adjust it for the difference in windspeed you have observed for your site and tower height, and prepare a wind distribution forecast for each month of the year. Let's say you have recorded wind speeds at a 40 foot height and find that winds at your site average 20 percent more than those recorded by the weather station's anemometer at 30 feet high.

Here's the way the wind distribution forecast for your site might look for a 30 day month:

Weather station wind (anemometer at 30 ft.)		20% adjustment for your site
Wind speed (in m.p.h.)	No. hours per month	Wind speed (in m.p.h.)
0— 7	286	0— 8.4
8	49	9.6
9	42	10.8
10	55	12.0
11	21	13.2
12	65	14.4
13	24	15.6
14	34	16.8
15	41	18.0
16	14	19.2
17	12	20.4
18	10	21.6
19	8	22.8
20	7	24.0
21	6	25.2
22	6	26.4
23	7	27.6
24 and above	33	29.0 and above

If you have a wind plant that starts generating electricity at 8½ m.p.h. or higher, you'd have 286 hours this month that no power could be generated. You need this information when it comes to sizing the battery bank to store electricity during calm periods. If the 286 hours are distributed through the month, you can get along on much less battery capacity than if the 286 hours of light winds occur in one stretch.

Each wind plant manufacturer computes an "output curve" for his machine. With the above figures on wind speed and the output curve for the plant you consider installing, it's a fairly easy exercise to forecast the kilowatt hours the equipment would produce for that month, at the site and height measured.

Notice that the table below is for example purposes only; the output of the generating plant at various wind speeds is taken from data developed by the manufacturer as an average of several tests with the equipment. Actual performance of any equipment may vary with the efficiency of the entire system. Assuming an overall system efficiency of 30 percent, here is the actual power—in watts—that might be expected from wind machines of various rotor diameters:

Rotor diameter (in feet)	Average wind speed (miles per hour)				
	5	10	15	20	25
	Watts of power generated				
4	2	19	64	152	296
6	5	43	144	341	665
8	10	76	256	605	1183
10	15	119	399	947	1848
12	21	170	575	1363	2661
14	30	232	782	1855	3623
16	38	307	1022	2423	4731
18	48	383	1293	3066	5988
20	59	473	1597	3785	7393

For example, the Sencenbaugh Model 1000-14 plant has a 12 foot diameter propeller and gear driven generator that cuts in at about seven miles per hour. Using the output curves for the machine, here's the way the Scenenbaugh might be expected to perform:

Wind speed at site (in m.p.h.)	Sencenbaugh 1000-14 output (in watts)	Hours per month	Kilowatt-hours produced
0— 8.4	Minimum	286	Negligible
9.6	160	49	6.7
10.8	200	42	8.4
12.0	300	55	16.5
13.2	380	21	8.0
14.4	420	65	26.4
15.6	500	24	12.0
16.8	620	34	21.0
18.0	720	41	29.5
19.2	900	14	12.6
20.4	990	12	11.9
21.6	1000	10	10.0
22.8	1000	8	8.0
24.0	1000	7	7.0
25.2	1200*	6	7.2
26.4	1200	6	7.2
27.8	1180	7	8.3
29.0 and above	1000**	33	33.0
		720	233.7

* Although the plant is rated at 1,000 watts output, the generator delivers up to about 1,200 watts at peak output.

** The 1000-14 has a rated wind speed of about 23 m.p.h. Power output declines rather sharply in winds above approximately 26 m.p.h. This kind of output curve is typical for wind generating equipment, although the rated output speed may vary from 15 to 30 m.p.h.

Again, you can see how electricity production increases with rotor size and, even more dramatically, with wind speed. With most equipment, and most wind distribution patterns, the biggest percentage of the power is produced by medium strength winds of 15 to 20 m.p.h.

What Type of Rotor?

Wind machine propellers, rotors, and turbines are classified in two categories: vertical axis and horizontal axis. Rotors with their axis or shaft parallel to the wind stream are horizontal rotors. With this type of propeller, a vane or some other means is needed to keep the rotor facing into the wind. The spinning propeller acts some-

what like a gyroscope, and does not track changing winds quickly.

Vertical axis rotors and turbines have their axis perpendicular to the wind flow (at right angles to the earth's surface). These rotors are "omni-positioned" to accept wind from any direction, which lets them make more use of the extra power in gusty, shifting winds.

Multi-blade turbines are those used most often with the familiar water pumping windmill. The rotor generally has 16 to 40 flat blades (or nearly flat blades) on a horizontal axis, and thus develops high starting torque, but low maximum speed and low end horsepower. The high starting torque makes this type of machine well suited for

Familiar water-pumping multi-blade turbine in the foreground is contrasted with the high-speed propeller that develops 200 kw near Clayton, N.M. (Courtesy of U.S. Dept. of Energy/NASA)

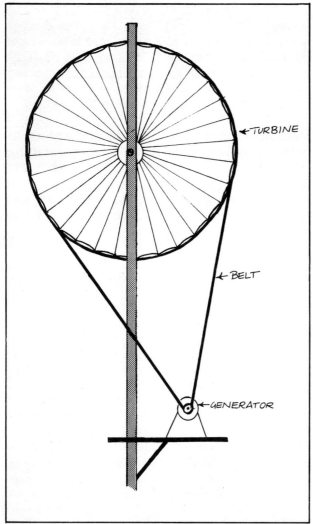

The super-speed turbine drives a generator with a belt, rather than through a shaft.

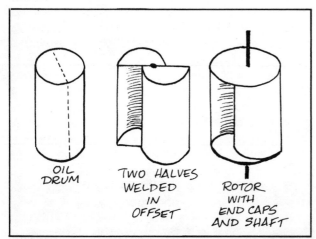

Savonius rotors are low-cost when built from a 55-gallon oil drum cut in half and welded in an offset manner. Finished rotor has end caps installed.

pumping water, as the turbine must start pumping on the first revolution. But, it's not particularly efficient for generating electricity.

High-speed propellers are made with two and three air-foil blades on a horizontal axis rotor. The propeller blades resemble those on an airplane, but are aerodynamically designed to grab energy from the wind, rather than propel an aircraft through the wind. Three bladed propellers are easier to balance and are most stable in shifting winds. This type of machine has low starting torque, but high horsepower at high r.p.m's. Both the high speed propeller and the multi-blade turbine require some means—a tail vane or rudder—to keep the rotor oriented into the wind.

Super-speed turbines resemble multi-blade turbines to the casual observer. But, these turbines have airfoil blades designed to grab and dump air quickly, so that "used" wind from one blade does not interfere with the blade behind. This type of turbine was first developed by William L. Hughes at Oklahoma State University, who also developed the idea of using a belt around the perimeter of the turbine to drive a generator directly, rather than through a shaft and gear box.

Savonius rotors are vertical axis machines with high starting torque but fairly slow top end speeds. Many Savonius-type rotors are homemade from oil drums, which are split to form the two halves of the rotor and offset as shown in the drawing. End caps are added to complete the rotor. This makes a low cost rotor, and several rotors can be stacked on the same shaft to increase wind surface area.

Darrieus rotors are also vertical axis machines, with two or three blades that somewhat resemble an egg beater turned upside-down. The machine has high efficiency, but very low starting torque. In fact, a Darrieus rotor alone is not self-starting. In wind plant systems, this rotor is used in combination with a small Savonius rotor to make it self-starting, or is started electrically by motorizing the rotor. The Darrieus rotor is always positioned to accept the wind, and rotors can be stacked for extra power and generating ability.

Getting up in the Air

As noted earlier, wind speed at 90 feet above the earth's surface may be 50 percent higher than at ground level, and is a lot steadier. A wind plant needs to be supported securely at a height where it has a clear wind-stream, free of turbulence and wind "shadows" caused by nearby buildings, trees, and other obstacles. A rule of thumb says the machine should be at least 30 feet higher

This modified Savonius rotor, built by Oklahoman Odell Morgan, features six barrel halves on a rotor, to power an automobile differential.

An auto air conditioner compressor is driven by the wind rotor, with 12-volt generator to produce power to engage the electric clutch on the compressor.

than the nearest obstacles within a 300 foot radius in all directions.

Generally, that means a tower—at least for larger wind plants. Smaller plants possibly can be mounted on well-guyed poles. Roofs and sawed-off trees make poor wind generator platforms. The elevated structure must withstand considerable lateral thrusts produced by high winds and must be fitted with a special adaptor platform for the particular model of wind plant to be mounted.

Most commercially made towers are built of galvanized steel, to resist corrosion. The tower needs to be well-anchored, usually on concrete pads or footings, and should be equipped with lightning protection. Needless to say, erecting a 50 foot tower and mounting a thousand pounds of equipment on it are undertakings that call for caution—and the right kind of equipment. Many wind plant dealers will not warranty their equipment unless the plant and tower are installed by qualified people; some distributors require that their own field supervisors oversee the installation.

Distributors and manufacturers of both wind and water powered generating equipment often will perform much of the engineering work associated with the site, either on location or from raw data sent by the homeowner.

SMALL-SCALE HYDROELECTRIC

Flowing water was an important energy resource for early American settlers, who tapped the mechanical energy in streams to mill grain, saw lumber, spin and weave cloth, and perform other essential functions. Modern day pioneers, who find themselves with a small running stream on their property, can harness this resource with equipment many times more efficient than the old water mills used by their forefathers.

Not many city dwellers have creeks and brooks coursing over their property. But many farms, homesteads, and other rural tracts afford small streams that could be put to work. There are many known potential sites for small hydroelectric systems around the U.S.; streams too small to interest the Army Corps of Engineers, but large enough to be good candidates for small scale power installations.

In a study done for President Carter, in connection with the National Energy Plan, the U.S. Army Corps of Engineers estimated that there is a potential power supply of 54.8 billion watts at *existing* dams in the U.S. Most of these are small, "low-head" (less than 65 feet high) dams that have been abandoned as power sites or never developed. Of about 50,000 existing dams with a power potential of 25 kw or less, only 2,000 are now generating electricity.

Power—some power—can be obtained from about any stream with a year round flow. However, the volume, gradient, and steadiness of the stream's flow will be important considerations for evaluating any site as an electricity producing resource. Even marginal sites may be improved by the installation of pump-back storage equipment. These hydroelectric setups use part of the power generated during off-peak periods to pump "used" water back into the storage impoundment.

Individual hydroelectric plants have been built with capacities as small as 500 watts' capacity. But the cost of engineering, building a dam and locating the hardware for such a small plant is, in most cases, not much less than developing a site to produce several times as much power. It's an economic situation similar to that of publishing this book. The expense of typesetting, making photographic plates, etc., is the same, whether the publisher prints 10 books or 10,000. However, the cost *per unit* is considerably less with the greater volume. Therefore, the economic facts of hydroelectric power generation make relatively larger plants a better paying proposition.

In many areas, there are cooperative or joint ownership possibilities; where one property owner may have a suitable site for hydroelectric power to serve several nearby homes or farms. Several homeowners could subscribe the capital needed to develop the site, install the machinery and distribution lines, and share in the power generated.

"I suspect that cooperative ownership of one or more largish plants would be difficult to administer to everyone's satisfaction," says Edmund Coffin, of Entertech, a manufacturer of wind power generation systems "Who gets priority on power use? Who pays for maintenance and repairs? What Federal Power Commission or state public utility commission rules would apply? I believe that establishing a local ownership utility might prove more practical and workable in most cases."

Coffin is talking about wind plants; but, the points he makes apply as well to hydroelectric systems. If two or more neighbors put up the money to build a plant on land owned by one of them, what happens if someone moves away, or if the neighbors have a falling out? Any such association probably should have the legal wrinkles ironed out by an attorney familiar with energy rules and regulations. But, the fact remains that sites exist in many areas for developing hydroelectric systems that would provide more power than would be needed by one household or one farm.

Evaluating the Site

The amount of power available at a potential generating site depends on the flow of water and the "head" or vertical distance in feet through which the water falls. In terms of horsepower, the available power can be expressed by this formula:

$$\text{h.p.} = \frac{V \times H \times 62.4}{33,000}$$

In this formula, V is the volume of water in cubic feet per minute; H is the head, or vertical distance, between the

Two- and three-bladed Darrieus rotors are high-speed machines often used to power generators. This 15-foot diameter (about 5 meters) rotor is built by Dynergy Corp.

Darrieus rotors can be "stacked" on the same shaft to provide more wind-facing area. This triple-decker turbine is installed at the Times Publishing Company, Clearwater, Florida. (Courtesy of Dynergy Corp.)

dam and the power plant; and 62.4 is the constant weight of a cubic foot of water in pounds. The product, of multiplying volume times head times 62.4 pounds, then is divided by 33,000, the number of foot-pounds per minute in one horsepower of work. The end result of this formula should be reduced by about 20 percent to allow for friction and slippage in the system.

Before you order a water turbine (called "wheel" or "runner") and generator, you'll need to do considerable engineering to determine:

1. Maximum flow of the stream.
2. Minimum flow.
3. Head or fall of the water.
4. Length of pipe (called "penstock") needed to get that head.
5. Water condition (clear, muddy, sandy, acid, alkaline).
6. Soil condition.
7. Minimum tailwater elevation (below the plant).
8. Area and depth of storage pond behind the dam.
9. Horizontal distance from dam to power plant.
10. Distance from power plant to point of use of the electricity.

And, as with any other private generating plant, you'll need to make a fairly accurate estimate of the amount of power needed and power consumption patterns—unless you'll be installing a plant with a capacity well in excess of anticipated usage.

Using Your "Head"

Head is the elevation or fall of water from the dam or catch basin to the powerhouse turbine. A high head, or long fall of water, produces the most power at least cost and at least consumption of water.

New turbine designs, such as Pelton type wheels, utilize the velocity of water directed through specially designed jets to get an amazing amount of work from relatively small turbines. To illustrate the importance of head in a hydroelectric plant, here are performance figures for the new six inch diameter Peltech wheel developed by Small Hydroelectric Systems and Equipment, of Arlington, Washington:

Major hydroelectric plants now produce about eleven percent of the U.S. power needs; smaller dams and sites around the country could provide another 55 million kw of power with the installation of small hydroelectric systems.

Head (which translates into water pressure) in a hydroelectric plant has almost the same kind of linear effect on power output that wind speed does in a wind power generating plant. You'll notice from the chart below that doubling the head from 50 to 100 feet results in a threefold increase in electrical power generated.

Head can be measured with either a surveyor's transit and poles, or with a hand level. Two people are generally required for this job; one to hold the pole, and one to sight through the transit or read the level. Sight through the transit to a point on the pole and measure the height, including the height from the ground to the transit in the initial measurement. Then move to the next lower elevation and repeat the process. Total all measurements taken to establish total head.

Low head systems, with less than 50 feet of fall, cannot make as good use of the highly efficient Pelton-type wheels. Turbines of the cross flow design, or water driven propellers are commonly used at sites with low water head. These require a much greater volume of water and develop less horsepower per cubic feet of flow than do high head systems.

Head (feet)	Water Flow (Cu. ft. per second)	Horse-power	Turbine Speed (in r.p.m.)	Watts Generated
20	.196	.3	643	187.5
40	.276	.8	910	500
50	.309	1.2	1018	750
80	.392	2.5	1286	1,000
100	.438	3.4	1440	2,125
200	.618	11.0	2040	6,875
300	.757	20.1	2490	12,562

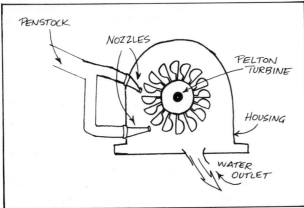

Efficient Pelton-type turbines require a high-pressure, high-velocity stream of water to strike the cups, turn the wheel and produce speed and power.

Measuring Flow

The other critical element in the water-power formula is water flow, or volume—usually measured in cubic feet per second or per minute. The potential power source at a site depends on the lowest and highest seasonal flow. The volume of water in a stream should be measured at different seasons to find both the peak and low flow periods.

There are several ways to measure stream flow. Here are the three most commonly used:

1. For a small stream, you can use a temporary dam and small container to estimate the flow. Divert the stream channel with a dam, so that the entire quantity of water can be caught and measured. You may want to install a large pipe in the temporary dam to draw off the water. Use a container of known volume and time the period it takes for the water to fill the bucket, barrel or other container. Divide the quantity of water (in gallons) by the number of seconds it takes to fill the container, then multiply that result by 60 to get gallons-per-minute of flow. Divide this figure by 7.5 to find the cubic feet per minute (A cubic foot of water contains about 7.5 gallons).

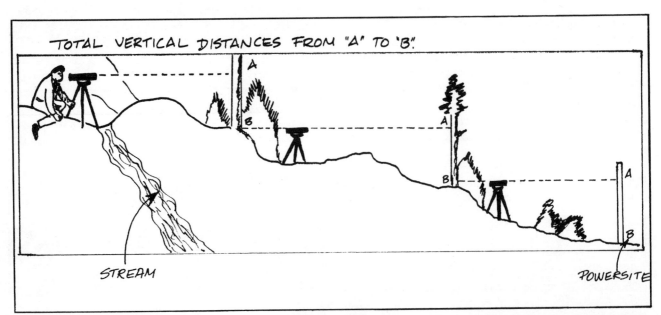

Measuring head, or vertical distance from water source to powerhouse.

High-head systems can produce a lot of power with a minimum of hardware. (Courtesy of Small Hydroelectric Systems & Equipment)

2. For a larger stream with a known type of bottom, you can use a float to get a rough estimate of the stream flow. Choose a windless day and a fairly straight section of the stream to do your measuring.

Measure a distance along the stream—at least 30 feet—and stretch a line tightly across the channel at the beginning and ending points. Time the period it takes for a float to pass between the two lines. Use a float that rides high in the water—you may want to attach a flag to make the float easier to keep in sight—and time the float several times to get an average rate of flow. Correct the reading you get by a factor of 0.8 for a stream with a smooth bed and banks, and by 0.6 for a rock strewn, hilly stream.

Next, determine the cross sectional area of the stream. Do this by multiplying the width of the creek by the average depth. Take the measurements in several areas along the stretch, and avoid making measurements in any pools or eddies. Average the readings.

To finally compute the flow, multiply the distance in feet the float traveled (with the correctional factor of 0.8 or 0.6 figured in) times 60. Divide by the number of seconds it takes the float to travel the measured distance. This gives the velocity in feet per minute. Then, multiply this velocity figure times the cross sectional area of the stream to get the flow in cubic feet per minute.

3. The weir method of measuring stream flow is the most accurate, but involves the most work, unless your stream already has a dam constructed. A weir is a dam with a level opening or slot about six times as wide as the greatest depth of the stream. The dam is installed and sealed so that all of the stream's water passes through the measured opening.

When all water is flowing through the weir, place a board or pole across the stream at some distance upstream from the weir dam. Then, place a second board across this and the dam, and level this board. Measure

Measuring the flow of a stream in cubic feet by, (A) cross-sectional method, and (B) by weir method.

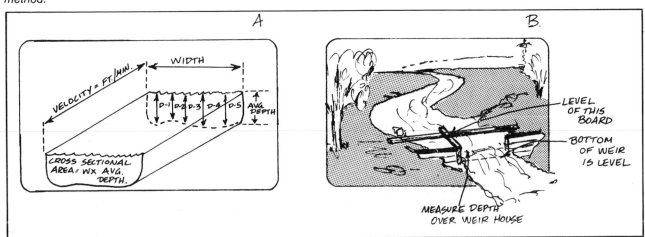

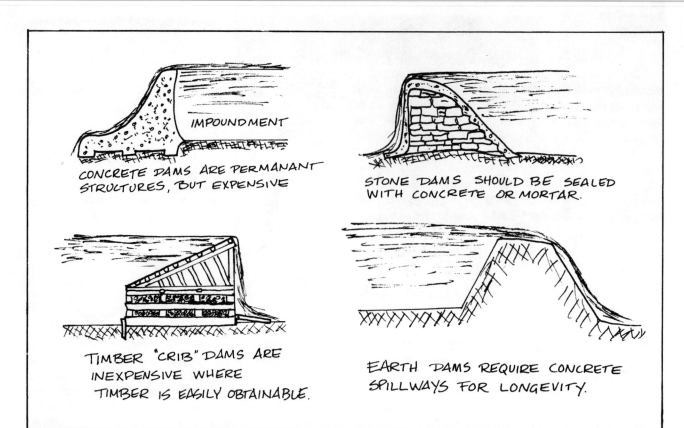

Dams for use with low-head power installations.

Depth of water over weir	Width of weir (inches)					
	12	18	24	30	36	42
	Flow (in cubic feet per minute)					
1 inch	4.8	7.2	9.6	12.0	14.4	16.8
2 inches	13.7	20.5	27.4	34.2	41.0	47.9
3 inches	25.1	37.6	50.2	62.7	75.2	87.8
4 inches	38.6	57.9	77.2	96.6	115.8	135.2

the depth of the water above the bottom edge of the weir and compute the flow from the table above.

Again, flow should be measured at different times of the year to get an idea of the power potential in all seasons. The measurement in seasons of lowest flow are particularly important in computing the amount of power that can be developed when the water resource is at its lowest point.

Building a Dam

With a few sites, where there is enough water to cover the intake of a pipe (penstock) or where part of the stream can be diverted into a catch basin or power channel, a dam may not be absolutely necessary with a hydroelectric plant. But in most cases, some impoundment of the stream will be needed to get a higher head than the stream itself offers, and to level out the supply of water to

the turbine. A dam also provides a settling basin where trash and water-borne material can settle out of water that is run through the turbine.

When you build a dam in a stream, a storage reservoir is created that can be used to good advantage to conserve the supply of water during times when the turbine is consuming more water than is flowing in the stream; or, to supply more water than the stream flows during periods of peak electrical demand. The load on any plant is seldom, if ever, fixed; it varies with the needs of the power consumer.

Dams can be built of earth, stone, concrete, timber, or combinations of materials, depending on the nature of the site and what materials are handiest and lowest in cost. Small streams can be dammed fairly easily, but larger streams, and streams in areas subject to much flooding, need professional engineering advice to prevent some costly disappointments.

Pump-back storage is a growing idea with hydroelectric systems. Pumped storage projects use the same principles as conventional plants, but have a tailwater pool below the power plant to catch water that has done its work in the turbine. Then, during off-peak periods—as at night, when household electrical loads are lightest—the water is pumped back into the upper reservoir, to be used again. So far, no one is claiming that pumped storage is at last the answer to the perpetual motion machine puzzle, but it can let you get a lot more electricity from each cubic foot of water.

Control Equipment

As with wind-powered generators, hydroelectric plants can be used to generate either AC or DC power. Generally, however, if a site will develop 1,500 watts or more, standard 60-cycle AC power is probably the best choice for most installations. For smaller sites, DC systems have the added advantage (although more cost) of storing power in batteries until it is needed.

With most hydroelectric plants, some kind of speed governing system is required—particularly if AC power is

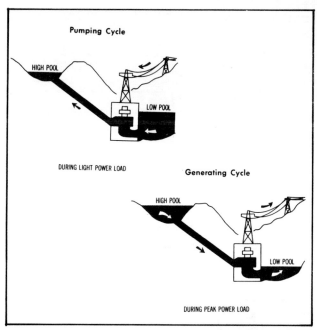

Water can be pumped back to the storage reservoir during light electrical load, then re-used to generate more power when demand increases. (Courtesy of U.S. Dept. of Energy)

Typical layout of a low-head small-scale hydro-plant, such as the Hoppes self-contained unit manufactured by the James Leffel & Company. (Courtesy of Nat. Academy of Sciences)

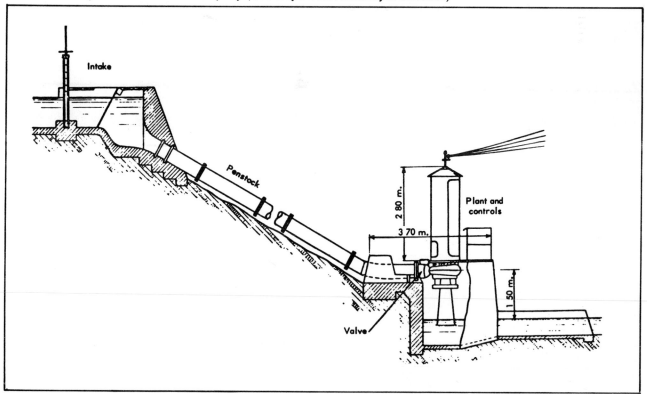

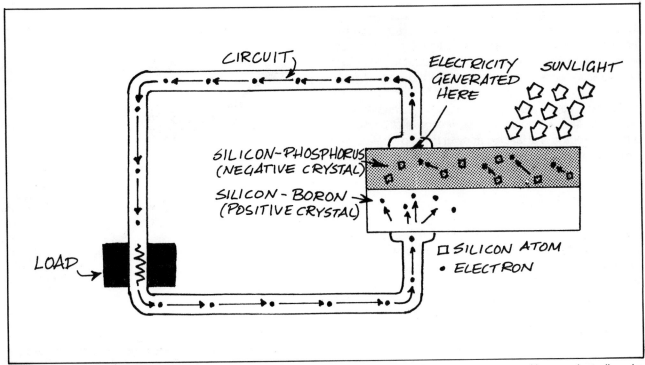

SILICON-PHOSPHORUS
(NEGATIVE CRYSTAL)

SILICON-BORON →
(POSITIVE CRYSTAL)

LOAD →

CIRCUIT

ELECTRICITY
GENERATED
HERE

SUNLIGHT

□ SILICON ATOM
• ELECTRON

How a solar cell works.

to be generated at level cycles. Woodward-type governors that work by centrifugal force are reliable, but costly. New solid state electronic load diversion systems work very well with AC generators. Overspeeding is no real hazard with these systems when they employ an "over-under" cycle switch and automatic bypass valve. The plant shuts down automatically if the AC cycles move out of a pre-set range. The bypass, or "blow-by" valve is held closed by a solenoid powered by the AC generator, and opens instantly if power is interrupted.

Hydroelectric plants can be hooked up to utility company grids in "co-generation" systems, through a synchronous inverter or with an AC induction generation system, as described earlier in this chapter. The principle is the same: once grid power is lost, the generator shuts down instantly.

DIRECT SOLAR ELECTRICITY—HOW SOON?

Electricity produced directly from sunlight has many benefits: it's inexhaustible; intermittently available everywhere; requires no fossil fuels that can be rationed by other nations; and can be harvested with equipment that has no moving parts to wear out.

It also has some rather obvious drawbacks. For one, the sun doesn't shine all the time. For another, you'd need

about 200 square feet of solar cells to generate 3,500 watts of electricity. But, the biggest thing that hampers direct conversion of sunlight to electrical power is *cost.*

The technology has been around since the 1950's to generate electricity directly from solar energy, through a photovoltaic effect that occurs when light hits certain sensitive materials and creates an electrical current. The most commonly used material in solar cells is silicon, a semiconductor. But ordinary silicon crystals don't work. The material must be processed to align the molecular structure of the silicon, and to add a few atoms of another material—such as boron or phosphorus—to the silicon atoms. This is an involved and costly process. Although the cost of solar cells has dropped in recent years, technology has not yet brought widespread use of direct solar within reach of many Americans.

By the mid-1980's? Maybe. The U.S. Department of Energy is embarked on an ambitious project to bring the price down to the point where photovoltaic cells can make a "substantial" contribution to America's energy needs by the year 2000.

A typical solar cell contains specially coated layers of silicon (or other light-sensitive material) with external wires attached. When light strikes the cell, electrons are released and an electric current results from the flow of these dislodged electrons. Individual cells are connected electrically to form solar "modules" or building blocks.

First village-size photovoltaic power project is the Papago Indian village at Schuchuli, Arizona. (Courtesy of U.S. Dept. of Energy/NASA)

For example, 40 of these cells could be connected together to provide enough electricity to charge a 12 volt automobile battery.

Solar cells generate DC power, which must be inverted to AC for most household use. However, the economies of scale appear to favor small installations. That is, there is not nearly as much essential hardware and engineering as with fossil fueled or nuclear power plants—solar cells can be ganged together to make whatever size generating plant is desired. This makes direct solar power as competitive for small scale systems as for larger generation stations, which means that using the sun as a source of electricity is catching on first in remote areas hard to serve with centrally-generated power.

In fact, the first village in the world to obtain all of its electricity from photovoltaic conversion of sunlight was Schuchuli, Arizona. Schuchuli (sometimes called "Gunsight") is a small Papago Indian village about 17 miles from the nearest available utility power.

On Dec. 16, 1978, the village threw the switch on a direct solar conversion system to pump water, power lights, run washing machines and refrigerators for residents. The Schuchuli installation was built and is being studied by the U.S. Department of Energy and the National Aeronautics and Space Administration agencies not as hard pressed to show a return on investment as are most homeowners.

Obviously, since the sun doesn't shine all the time, direct generation of electricity from sunlight requires batteries or some other system to store the energy produced. That may not be much of a problem with small scale installation, where DC power can be stored in batteries, then inverted to AC current as needed. But, it could interfere with the federal government's goal to build giant megawatt solar cell systems by 1990 to provide power to whole towns and utilities. Unless new ways are found to store electrical energy, some gigantic batteries will need to be built in the next 10 years. Otherwise, conventional generating equipment will be needed to pick up the load at night and on cloudy days—which means that conventional plants still will be needed, at least for standby capacity.

Layout of a solar-powered Rankine engine electricity generating plant.

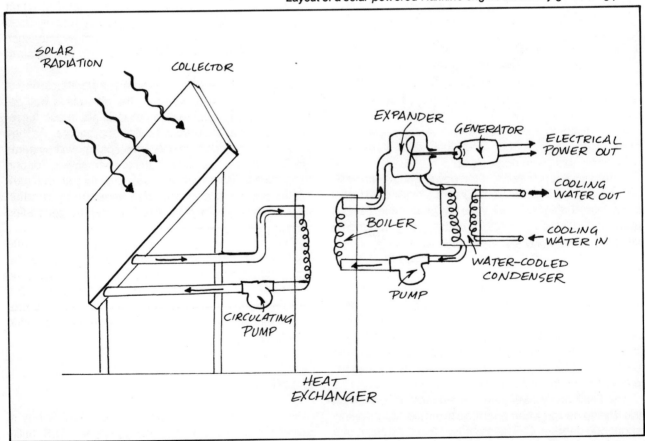

Buck Rogers on the Line

In a conventional power plant, heat generated from burning fossil fuel is converted to electric power, using a steam turbine or other type of heat engine. We've already noted that converting thermal energy, to mechanical energy and then to electrical energy, is not particularly efficient, unless some use can be made of the surplus heat.

New ways are being developed to use low grade heat from a variety of sources—waste heat from power generation and industrial processing, solar energy, geothermal energy, burning wood and other solid fuels—to generate electricity. Actually, most of these "new" systems are old ideas that have been dusted off and put to work.

The Burlington Electric Department, Burlington, Vermont, has converted a 10 megawatt coal fired plant to use wood chips, and shaved a penny off the kw-hr cost of electrical power. The company now is building a 50 megawatt wood fired plant from the ground up that will run on 1,500 tons of chipped wood wastes per day.

Sun Power Systems, Inc., of Miami, Florida, has come up with a solar heat engine that uses circulating freon to produce up to 15 kw of electricity; then uses the cooling water from the engine to pre-heat water for a conventionally fired boiler.

The Agricultural Engineering Department of the University of Missouri has a self-sufficient hog farm model that digests swine manure to produce methane gas. The gas powers an internal combustion engine to drive a 15 kw AC generator. Heat from the engine's cooling water and exhaust is piped to hog buildings to heat the pig's living quarters. The electricity produced powers lights, feed, handling equipment, water pumps, and other electrical devices.

Much research is being done on fuel cells, which generate electricity by electro-chemical action from fuels such as alcohol and methane gas.

Perhaps most promising of the power generation vehicles now on the horizon are the various types of external combustion engines that can use low grade heat. Among these are updated versions of the Stirling engine, first developed in 1816 by Robert Stirling, a Scottish minister—further evidence that there isn't much that is brand new in the energy field.

These engines function in much the same way that internal combustion engines work; except, that the heat source is outside the cylinder or expansion chamber. The heat source applies thermal energy to one end of the cylinder; cool atmospheric air is drawn into the cylinder and compressed by the piston; the air heats rapidly and expands, forcing the piston back on a power stroke.

The Rankine cycle engine is even more efficient than the Stirling design when operated at temperatures below about 300 degrees C. The most traditional example of a

The venerable steam engine upon which this scale model is based, may have potential as future power generating equipment—when fueled with wood, methane gas or alcohol.

Rankine engine is the familiar steam engine (which itself is enjoying something of a revival in some quarters); but, the Rankine uses organic materials instead of water (steam) as the working fluid. The most commonly used working fluids in Rankine engines are refrigerants, such as freon, which have a low vaporization (expansion) temperature.

At present, there are few commercial manufacturers of these "heat" engines. However, the principle is well established and several working models are in use for research projects and special service applications. As they come into more common use, small scale power systems may be the best use made of these alternatives. For one thing, they tie in well with a total energy system that generates electric power and also uses the by-product heat—which is normally wasted in a central generating plant.

These machines to generate electricity from low-grade heat and natural fuel sources offer a lot of promise for the future. Some of them, such as the Stirling-type engine, no doubt will be commercially available by the 1990's. But, for most homeowners who are looking for ways to produce their own "juice" right now, this hardware has limited application.

EMERGENCY ELECTRICITY

Whether you produce your own electricity or buy it, power brownouts and blackouts do happen. U.S. utility companies have a good service record—the average family is without power less than 24 hours in 10 years—but power outages are not unheard of, with both centrally generated and homemade power.

A loss of electricity can be costly, as well as inconvenient. Most power outages occur during—in fact, are caused by—disagreeable weather. A hurricane knocks down power lines and the accompanying rains flood your basement because your sump pump doesn't work. A summer tornado knocks down power lines, and the hot, humid weather in its wake causes a freezer full of food to spoil. Winter ice breaks wires, and water pipes freeze and burst.

Some emergency source of power may be needed by many homeowners to run those essential electrical appliances such as pumps, furnace blowers and refrigerator/freezers. In homes where ill or handicapped persons depend on electricity to power health support equipment, an emergency back-up is vital.

For most of us, the best insurance is a small gasoline powered auxiliary generator. A 3,000 to 4,000 watt generator will operate lights, furnace fan, refrigerator, sump pump and a radio or TV—although perhaps not all of these devices at the same time. You'll recall that the locked-rotor amperage of AC motors requires about three times as much current to start as the motor draws while running, even under full load. That means you'll have to allow for the starting current of those motors that switch on automatically, such as compressor motors on refrigerators and blowers on forced-air furnaces. Some emergency generators have a "surge" capacity over their rated running wattage. For example, a 3,500-watt generator may be running at two-thirds load and still have enough capacity to start and run a ½-horsepower motor. But you'll need to determine your electrical needs pretty closely to size an emergency generator accurately.

Don't buy more generator capacity than you absolutely need. After all, it is an emergency piece of equipment, meant to power only necessary electrical lights and devices until normal power is restored. But you should choose an emergency unit with enough capacity to run the equipment and appliances that are important to your family's health and comfort.

The generator should be versatile, portable and easy to operate and service. Purchase the unit from a local dealer who will provide reliable repair service. Figure on spending about $250 per 1000 watts capacity for an emergency generator in the 2,500 to 5,000 watt range. Reliable equipment doesn't come cheap. It's not an inexpensive way to generate electricity, from an operating standpoint, either. A 5,000 watt emergency generator will drink about a gallon of gasoline each hour it runs under load. With gasoline at over a dollar per gallon, that's a kw-hr cost of over 25 cents. But there is another advantage to owning a portable generator: you can take your electricity with you wherever you go, to power lights, tools and other equipment.

Incidentally, if you own a farm tractor, you may want to consider a generator, such as the Winco 15 kw model, that operates off the power-takeoff (PTO) shaft of the tractor. There's no separate engine to buy and maintain. You'll need about five PTO horsepower for each 1,000 watts of generator capacity.

In summary, there's no longer any *cheap* source of electricity. The days of the 1.5 to 2 cents per kw-hr for power from utilities are gone forever. The bulk of America's commercial power is generated by fossil fuels; each price hike in fuel means a boost in the cost of centrally generated electricity.

That cost price spiral has pushed electric rates to the point where many homeowners think seriously about generating at least part of their power needs. But, even if you have a good source of generating power—wind, water, or whatever—homemade electricity is not cheap, either.

Put a careful pencil to the economics of buying and installing generating equipment. If the move will not quite justify the initial capital cost right now, up-date your figures each time there's a hike in your utility rates. More and more homeowners who have generating plants are seeing their investment pass the break even point and move into the black ink side of the ledger.

9
CONTROLLING THE SYSTEMS

"Alternate energy should be enjoyed; not merely endured."

William Enter
Anabil Enterprizes, Inc.
Mustang, Oklahoma

A few natural energy purists seem to believe that it would be insincere to use electrical pumps, fans and other equipment in conjunction with solar, wood burning or other alternate energy systems.

However, the small amount of electricity consumed to open and close valves, dampers and run a few pieces of equipment in a system can be a good investment in comfort, convenience and—in the long run—economy. It lets a homeowner get a lot more mileage out of the energy produced by solar collectors or wood burning equipment. And, with electricity generating set ups, some method of "taming" the power output is essential to the satisfactory operation of the equipment.

If you will use pumps and fans to move heated air or water, it follows that you should invest a little more money and a few more watt hours in control equipment which allows the system to function more or less automatically. Tending solar collector to storage pumps, or manually adjusting the air inlet damper on a wood stove may not be big chores, but at times they can become a drag.

In fact, a variety of reliable *non-electrical* control devices can be put to work. Bimetal strips can control stove dampers. Ram-type heat motors can operate dampers, louvers and windows with no external source of power.

Even with electrical controls, new developments in solid state low voltage controllers and relays can free a homeowner from constant attention to his heating equipment—and do a better job of maintaining an even temperature than manual controls. Multi-input, multi-output programmed switches can, for example, be installed in solar heating systems to control: collector to storage pumps or fans, collector to space heat fans and switchover dampers, domestic hot water pumps; and then shift automatically to back-up heating systems (heat pump, oil furnace, electric baseboard heaters, etc.).

"An active solar system that uses a controller adjusts to weather conditions automatically," says William Enter, a pioneer inventor in the area of heating system controllers. "It measures the collector temperature in a solar heating system, measures the storage temperature and uses the information to profitably harvest solar energy. The controller also decides when it is *not* profitable to harvest solar energy and shuts the system down. A good controller properly installed can do a better job of sensing the weather and operating a system than a conscientious person who was on the job constantly."

Because of limited space here, and the great variety of control equipment on the market, we'll discuss controllers only in a general way. Many manufacturers of alternate (natural) energy equipment incorporate controls in the hardware package; others specify or recommend differential thermostats, switches, relays and other control equipment to be used.

NON-ELECTRIC HEATING/COOLING CONTROLS

The most familiar non-electric control equipment is the bimetallic strip or coil. This kind of device has two strips of dissimilar metals sandwiched together. One metal expands more rapidly than the other, which causes the strip to bend or the coil to flex, when heated and cooled.

Bimetal controls are used most frequently to control wood burning equipment draft dampers, but can be put to work in other space heating systems. Bimetal thermostats are standard equipment on several wood stoves, and most wood burners with controlled combustion can be equipped with this type of controller.

Another popular non-electric control with wide application is the expanding fluid type ram or heat motor that often is used to open and close shutters, windows, louvers or valves. These controllers feature a long pipe-like cylinder filled with thermally sensitive gas or liquid. As the fluid is heated, it expands to force a ram or piston along the cylinder. These controllers operate with a positive relation to the rise and fall of temperature, and most often are used to power ventilating equipment. For example, a

ram-type heat motor could be installed to open a bank of greenhouse windows when the interior temperature rises, then would automatically close the windows as the space cools down.

ELECTRICAL CONTROLLERS

The most universal heating system controller is the familiar wall thermostat, used with both conventional and natural heating systems. Thermostats utilize several methods of sensing and responding to temperature change (including the bimetal coil described above), but all perform essentially the same kind of function. When the sensed temperature falls below a pre-set mark, a set of contact points on the thermostat close, energizing a valve, pump, fan or other heating system component. The heating equipment operates until the temperature sensed by the thermostat is above the shut off setting (thermostats typically operate in a "start-stop" range of a few degrees) when the thermostat contacts open to stop the flow of electricity to the component.

Thermostats to control air conditioning equipment operate the same way, but in reverse. The contact points close at a high limit temperature and operate the equip-

Thermostatic Stove Control

Two "D" cell batteries, a thermosat and the electric motor from a toy make this controller for a wood stove damper. (Courtesy of Alternate Sources of Energy)

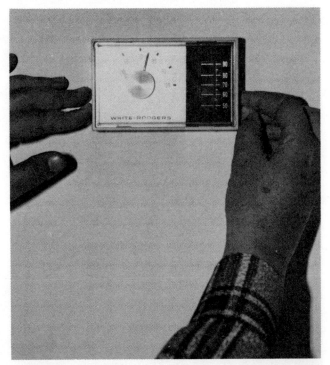

A single stage wall thermostat is the most universal heating and cooling system control.

ment until the sensed temperature is once more below the cut off setting.

Actually, thermostats designed to control the heat to living spaces are often two stage controllers. The first stage operates a blower or pump to circulate heated air or fluid from the furnace or solar storage to the space; the second stage operates the burner or other heat source. Or, the thermostat may start the heating system (open a gas valve, for example) while a second thermostat in the warm air plenum energizes the blower.

Controllers usually operate on voltages lower than the line voltage which powers the equipment they control. Thermostats typically utilize 12 to 32 volt electricity, then energize the pump, fan, heating coil or other equipment through relays. In other words, when a contact on the thermostat closes, a low voltage circuit is energized. This current operates an electrical relay, with much heavier contact points, to energize the 115 volt or 230 volt circuit which powers the heating equipment itself. The relay has heavier duty contacts; higher voltages would damage the thermostat.

"Differential" thermostats are often used to control solar heating systems. Since the solar system must depend on the outside weather, the controller used must read not only indoor temperatures (as a conventional thermostat does) but also read temperatures in the solar collector and heat storage. The most basic solar system would have a temperature sensor in the solar collector and another in the heat storage (rock or water) bin or tank. The collector fan or pump should operate only when the collected heat exceeds the temperature in storage. In most set ups, the pump or blower is operated only when collector temperature is 10 to 15 degrees higher than storage temperature.

For self-draining solar collectors that use water as a heat transfer and storage medium, the controller should

Operations and monitoring center between solar collectors and hot water storage tank is this Grumman Sunstream® heat exchanger module. (Courtesy of Grumman Energy Systems)

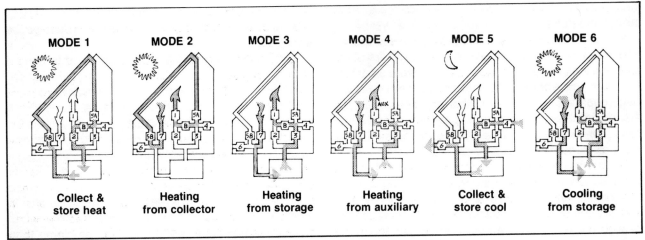

MODE 1	MODE 2	MODE 3	MODE 4	MODE 5	MODE 6
Collect & store heat	Heating from collector	Heating from storage	Heating from auxiliary	Collect & store cool	Cooling from storage

Motorized dampers can be grouped to handle heating and cooling air with one blower. In these drawings, dampers are numbered and blower is marked "B"; air flow is indicated by shaded patterns. (Courtesy of Heliotroipe General)

allow some collector "overheating" before the pump is signalled to start. This delay avoids a premature shut-down, or "short-cycling" of the system when the first cool water from the piping enters the collector and momentarily cools it down by several degrees.

In a more sophisticated differential controller, the above functions may be augmented by controller modes to operate fans, dampers, and pumps in other parts of the system. For example, a thermostat in the living space sends a signal to the controller for more heat. The controller decides whether the heat can most efficiently be supplied by (a) the collector itself, (b) stored heat, or (c) one or more back up conventional heating systems.

Heating system controllers, if chosen to fit the job to be done and installed correctly, should be virtually trouble free. However, there is a tendency among some builders to over-design controllers to the point of mere gadgetry. There's no good reason to install more control instruments than are needed to perform the basic functions with a degree of convenience and reliability. The most workable systems are generally the easiest.

Some natural heating systems will require "fail safe" and safety controls, too. These may or may not be electrically operated. But, the pressure and temperature relief valves in hot water systems (both for domestic hot water and for space heating) probably should *not* be electrically operated, since these devices may need to operate to prevent excess pressure, or temperature, in the system in a power failure. Also, some temperature limiting device should be installed in solar heated and most wood heated domestic hot water systems to prevent tap water from becoming hot enough to scald flesh. An automatic mixing valve at the faucet is one way to do this.

It's a good idea to install remote thermometers at key points in a solar system to keep a check on what the controller is doing. A thermometer in the collector, another in the storage tank or bin, and perhaps a third in a domestic hot water heat exchanger can let you keep track of what's going on. To keep even closer tabs of what the system is doing, you may want to install thermometers at the inlet and outlet of each of these components. Also, many controller panels are equipped with indicator lights that tell you at a glance what mode of operation the system is in.

A conventional back-up heating system (oil, gas, electric) usually operates in tandem with the natural energy heating system. It's a fairly simple matter to wire the control circuitry so that the back-up automatically goes into service when the natural source of heat fails to carry the load. Dual fuel and "piggyback" wood burning equipment typically has this type of controller. When the thermostat calls for more heat to living spaces and the burning wood can no longer meet the demand, the gas or oil burner comes on. This chain action control can be handled by one two-stage thermostat. Or, in some set ups, two separate thermostats may be installed: one to control the combustion air on the wood burner and one to energize the conventional fuel burner. Where two thermostats are used, the unit that controls the conventional burner is set a few degrees below the one that controls the wood burner's damper.

Conventional back-ups to solar heating systems are controlled in much the same way, with one additional function. A damper or some other cut-out usually is installed to prevent the conventional heater from heating water in the storage tank or rock in the storage bin.

Each type of heating plant—natural, conventional or both—requires special features in its control system. Even the simplest controls should direct the major routine functions of the system and include high-limit controls to prevent overheating or over-pressurizing the system.

Because each system's requirement can be different from the next, it's a good idea to go over the performance specifications of the manufacturer for each component in a system. For example, pumps, fans, and other electric motors should be started with relays that are heavy enough to handle the surge of starting current. Relays carry a motor rating. If a homeowner has the knowledge and experience to lay out and install the control system, he's probably familiar with the finer points of sizing and matching controller elements. But, if there's any doubt about which kind of controller to install or how to install it, the services of a qualified technician can avoid future problems.

Controllers, relays and other control equipment should be located where they will have convenient access for inspection and servicing. With some planning, you can save money on wiring the system. For example, output relays (that throw the 115 volt switches to energize fans or pumps) can be located near the equipment. This allows low voltage wiring for the longer part of the run; with heavier wire a shorter run from the relay to the equipment.

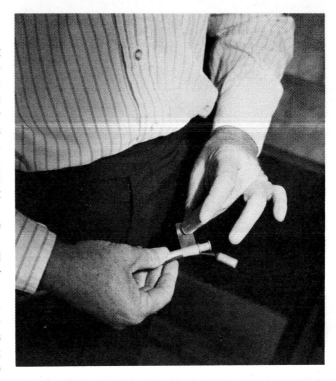

GENERATING PLANT CONTROLS

As noted in Chapter 8, you'll need some method to "tame" the power generated by wind, water, and other natural sources.

In a battery storage system, where a direct current generator is essentially a battery charger, a voltage regulator is needed to assure that battery cells do not become overcharged. Overcharging can literally destroy a battery. The regulator limits the charge voltage, and reduces the current through the battery as cells near the peak charge. Also, a voltage regulator allows the DC generator to maintain steady voltage as the turbine speeds up or slows down.

AC generators (called "alternators") must be run at a steady speed, to produce the 60 hertz (alternating cycle) power. For example, a four pole alternator, driven by a water turbine, must run at exactly 1,800 r.p.m. to deliver power at standard voltage and frequency. (The number of "poles" determines the alternator's speed: a two pole unit would need to run at 3,600 r.p.m.; a six pole alternator, at 1,200 r.p.m.). Both mechanical and electronic speed governing controllers are built to be used with several types of generating plants. Electronic controllers are typically "power balancing" or load diverting devices that sense power demand and signal more, or less, power from the turbine, engine, or other power source.

Remote thermometers come in handy for spot checking temperatures in different parts of a heating system. Here, the temperature sensor is taped to a fishing pole and "probes" the attic temperature.

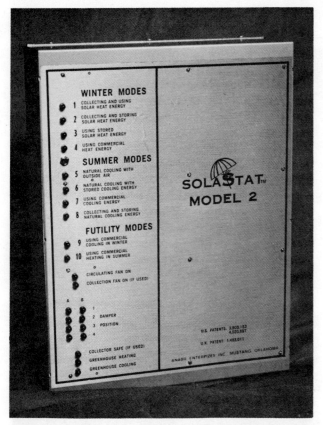

Some contollers have small indicator lights to show the mode of operation. (Courtesy of Anabil Enterprizes, Inc.)

Output governing equipment should be designed for the particular generating system being installed. Manufacturers and distributors of wind and water powered generators are the best source of information on the compatible controls to install.

Most household appliances are designed to use 60 cycle AC power, either 115 volt or 230 volt. With generating systems that produce DC power to charge a bank of batteries, the direct current must be converted to 60 hertz (cycle) AC power needed for these appliances. The equipment needed for this function is called an "inverter." ("Rectifiers" operate in just the opposite way, to flatten out the sine wave of AC power and produce a DC-like current).

There are actually four types of inverters; three of them commonly used to convert DC power to the voltage and wave form frequency needed.

Rotary inverters are battery powered alternators that produce 115 volt AC power. A DC motor turns a shaft which drives an alternator. Rotary inverters are not especially efficient—perhaps 60 percent on the average—but they do impose good voltage and frequency control. They are available to about 2,000 watts output, but cannot

Hydroelectric plants that produce AC power require speed and voltage limiting controls. (Courtesy of Small Hydroelectric Systems)

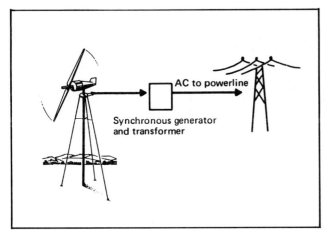

Wind generating plants can be tied directly to utility power grids through a synchronous inverter. (Courtesy of National Science Foundation)

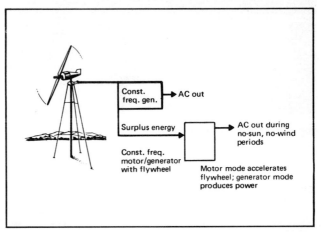

Wind energy also may be stored as mechanical (or "kinetic") energy in the motion of a flywheel. (Courtesy of National Science Foundation)

handle "surge" loads such as fan motors, refrigerator compressor motors, etc.

Vibrator inverters are more efficient than the mechanical rotary devices. In this type inverter, DC current drives a small vibrator assembly that imposes the sine wave to convert the power to 60 hertz AC. Vibrator inverters are probably the most economical to use in low power systems. However, they are limited to about 500 watts continuous load.

Electronic inverters use solid state semi-conductor technology to invert DC to AC in high power systems with good efficiency. These controllers are fairly expensive compared with other inverters; costing up to $2 per watt capacity.

Synchronous inverters, as described in Chapter 8, let all the power generated by the plant match utility power at standard line voltages and frequencies. This controller is the best choice for power plants installed where commercial grid power is available. The inverter "reads" the voltage and frequency of the utility power grid and keeps the generator's output synchronized with it. Solid state circuits in the inverter constantly adjust the varying output of the generator to match the line power.

These inverters have other important built-in controls. If commercial power is interrupted, two relays in the synchronous inverter open. The first relay disconnects the tie-in circuit with the utility grid power, so that none of the plant's output can be fed into the power line. The second relay disconnects the windplant, or other generator, from the inverter by sending the generator the same kind

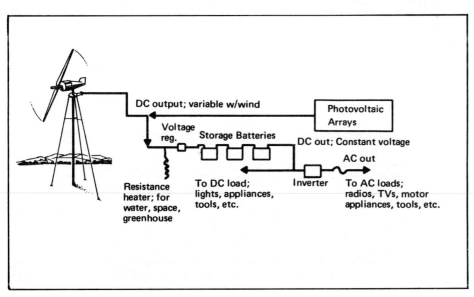

Wind plants that generate DC power usually are battery charging systems, and can be combined with direct solar power generation (Courtesy of National Science Foundation)

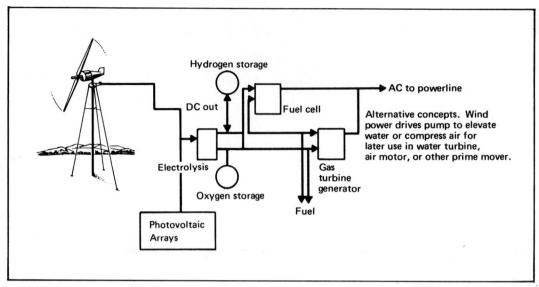

Wind generated electricity can be used to produce hydrogen and oxygen from water, through the process of electrolytic dissociation of those elements. The hydrogen can be stored and used as a fuel. (Courtesy of National Science Foundation)

of "shut down" signal it would get from a fully charged battery.

On a safety note: in installations where part or all of the circuits are wired for DC power direct from batteries, special fuses and circuit breakers will be needed. Direct current can "jump" a much wider gap than will AC power. A standard AC ribbon fuse will "blow" on a DC overload, but the current may arc across the burned area. Fuses and circuit breakers designed for DC current allow wider distances to prevent this from happening.

A couple of generations of Americans have become accustomed to the convenience and comfort of automatic controls. There's no reason to sacrifice these benefits to go to another source of heat or power.

But, plan your control devices and circuitry with at least as much care as you devote to other elements of a natural energy system. You'll probably want to install manual override switches for key functions, even if you completely automate the system—just in case the "nerve center" has a nervous breakdown.

10
HOMEOWNERS WHO ARE DOING IT

"Example is not the main thing in influencing others–it is the only thing."

Dr. Albert Schweitzer

Despite the fact that the lion's share of federal grant money for alternate energy research goes to major corporations, rather than to individuals and smaller companies, a good many of the immediately "borrowable" ideas right now are being developed by homeowners, tinkerers and small firms around the country.

When the natural energy industry grows up enough to be able to recognize its Henry Fords and Robert Fultons, they will probably be the Steve Baers and Andy Davises of today; homeowners who adopted alternate/natural energy ideas, made them work and then spread the word to others. These are the people who are taking the early licks in designing and building economical, workable natural energy systems. Not everything they try works perfectly the first time, of course, but that is true of pioneers in any field. Thomas Edison tried hundreds of filament wires before he discovered how to make a tungsten light bulb.

Alternate energy is a wide-open, fast-growing field of endeavor with plenty of room for a man (or woman) with a good idea. It's interesting to note how many individuals, who have come up with workable natural energy systems, have turned their experience into prosperous businesses in the past few years:

Steve Baer, who developed the "drum wall" idea of passive solar energy collection and storage, now has a thriving solar energy business in New Mexico. Henry Clews, who balked at paying a utility company thousands of dollars to bring power to his Maine homestead—and instead built a wind powered generating plant, now is with a major manufacturer of wind plants in Vermont. Andy Davis, (of Illinois) first got the idea for a totally underground home while exploring an old mine tunnel in Arkansas, now operates "Davis Caves, Inc.", a success-ful company that designs and constructs underground dwellings.

While their companies aren't bucking for the Fortune 500 listing just yet, these individuals and dozens more are developing good ideas into commercial business ventures. Of course, not all homeowners who come up with workable, money saving natural energy systems go commercial with their ideas. Here is a mixed bag of case histories of homeowners who are making natural energy work:

GOING UNDERGROUND

Winter winds blow hard across the Illinois prairie in Tazewell County. After watching winter fuel bills go past $150 per month to heat their above ground house, Andy and Margaret Davis decided to build a home snuggled into the good Illinois earth.

"Actually, I got the idea for an underground home a few years earlier, while in an abandoned mining tunnel in Arkansas," recalls Andy Davis. "It was hot outside, but the farther back I went into the mine, the cooler it was. At the time, I thought this would be a great place to live."

After that expensive winter above the ground, Davis went shopping for a south facing hillside site. He drew up plans to fit the site, dug a semi-circle into the hillside and built the footings, walls, floor and roof of reinforced concrete and stone. The result is; an eight sided, 1,200 square foot home with earth banked against all sides except the front (south); and with four feet of earth over the roof, where the family grows a small garden.

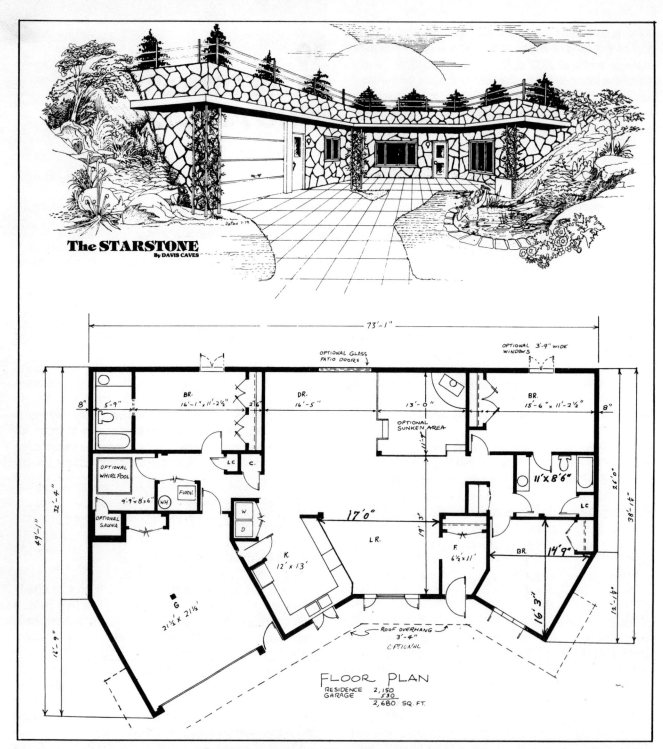

The STARSTONE
By DAVIS CAVES

FLOOR PLAN
RESIDENCE 2,150
GARAGE 530
2,680 SQ. FT.

Experience gained in building underground homes has been put to work in a commercial business by Andy Davis (Courtesy of Davis Caves, Inc.)

Davis estimates his total cost at $15,000, including the cost of the land, a septic tank and the home's furnishings—in terms of 1977 dollars and prices. The Davises did all the labor on the house, partly to save on construction costs, but mainly because Andy wanted to develop the techniques that are employed in building his unique "cave" type dwelling.

"Our homes can be built for the same cost—in many cases five to 10 percent less—as conventional housing," says Margaret Davis. "But the big savings begin after the home is built, in the form of reduced maintenance and greatly reduced heating and cooling costs."

For the first two winters in their underground home, the Davises heated their dwelling with 2½ cords of wood per year (harvested with about $1.30 worth of chainsaw fuel), burned in a single wood stove. Even when outside temperatures fall to well below zero, the home's interior stays at a cozy 70 degrees F.

Exterior walls are eight inch thick concrete; ceilings are one solid 10 inch thick piece of reinforced concrete, designed to hold 790 pounds of weight per square foot.

"This is an important feature," says Mrs. Davis. "We never use less than three feet of earth cover on top of the homes we build. Four, five or even six feet is better still, but the structure must be designed to withstand that kind of pressure."

Ground water is carried away from the Davis house through drainage tiles; the site is graded to divert surface water from the structure. The Davises now build their "Earth-Powered Homes" in many areas of the U.S. through a network of franchised dealers.

HOUSE IN THE BIG WOODS

When the author's family built a new tri-level frame house on a high ridge in the Missouri Ozarks, two key features were emphasized: (1) the house must be located to command a 12 mile unimpaired vista of the folding hills below, and (2) the heating system would make use of the most abundant natural resource—wood—but would not require night and day attention.

With some planning, both considerations were fairly easy to accommodate. The property included the highest elevation for miles around, thus a constantly changing scenic view to the south and west; and contained 20 odd acres of oak, hickory and ash hardwood trees—a perpetual supply of heating fuel.

The Ritchies built their home with R-30 insulation in the ceiling and about R-14 in the walls. The house was

Home of the Davis family is this rock and concrete structure snuggled into an Illinois hillside. The family grows a small food garden on top of the house. (Courtesy of Davis Caves, Inc.)

A single wood stove and 2½ cords of wood adequately heat the Davis underground home. (Courtesy of Davis Caves, Inc.)

Underground homes with one side at ground level are not greatly different in appearance–inside–from conventional homes. (Courtesy of Davis Caves, Inc.)

A dual-fuel (wood plus propane) furnace is the main heating system in the Ritchie home. Furnace takes 5 foot long logs.

oriented to turn the narrow dimension to the north wind's blast, and to provide a walkout basement on the sunny south side. No windows or other openings were made in the north wall.

A combination wood-propane forced air furnace is the principal heating unit for the three bedroom, two bath house. The furnace, built by Longwood Furnace Company, features a five foot long wood burning combustion chamber. In all but the coldest weather, a charge of wood will last 10 to 12 hours. The furnace is equipped with an auxiliary propane gas burner that automatically takes over space heating chores when the wood fire dies down. The whole apparatus is controlled by a central thermostat.

The furnace's gas burner rarely operates, except when the family is away from home long enough for the wood fire to die down. The L.P. gas also heats domestic hot water and does part of the cooking. With propane costs at about 55 cents per gallon, the year's highest gas bill runs less than $40 per month. The Ritchies use about 4½ cords of wood each year; that's less than a week's worth of not-too-frantic timber work. Wood is cut, split and allowed to air dry for at least six months before the heating season.

Vertical solar collector on a wall of the south-facing, walk out basement picks up extra heat on sunny days. Collector is framed with 2x4's and covered with two layers of polyethylene film.

Mason's earth-contact solar heated home crouches into a hillside in the Missouri Ozarks. (Photo: Bill Mason)

Part of the wood is burned in a Heatilator® heat circulating fireplace with circulating blowers. This unit, while not as frugal with firewood as the furnace, easily heats the 1,800 square foot home until temperatures fall below about 25 degrees F.

Some additional heat is supplied to the basement on sunny days by a small vertical solar collector mounted on the exterior south wall of the basement. The collector has about 80 square feet of area, and is baffled horizontally. A small floor fan pushes air through the collector and into the basement, where the heat is stored in the walls and concrete floor.

EARTH-CONTACT PLUS SOLAR
(And a Wood Stove for Good Measure)

When photographer Bill Mason and his artist wife, Virginia, planned their new home and studio, they incorporated several ideas to save energy and money. They

Mason's solar heated air handling system is based on this pattern, designed by U.S. Dept. of Agriculture engineers. (SEA-ARS, Clemson Univeristy)

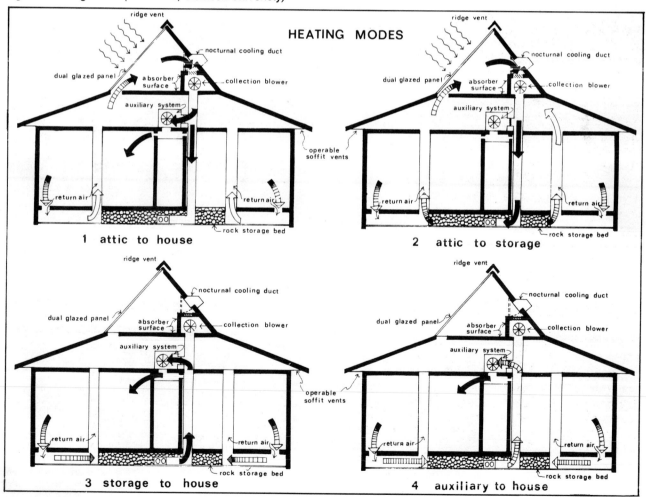

Heat storage bin holds 20 tons
of washed creek rock.
(Photo: Bill Mason)

Heated air from solar collectors pre-heat domestic hot water in these three 62-gallon tanks.
(Photo: Bill Mason)

built an earth-contact structure of CCA-treated wood, constructed a 400 square foot solar collector on the south-facing roof and installed a Franklin style wood burner in the living room. For insurance, Mason added a heat pump and 20 kw electric furnace as back up systems.

"I think some of the features we built in grew out of Bill's curiosity, as much as out of a desire to save energy," says Virginia Mason. "But we were snug during our first winter in the home (1978-79) and it was a record setter for cold."

Mason started by scooping an eight feet deep hole into a south hillside, then leveled and packed a foot of gravel in the bottom of the excavation. Construction started with a 12 by 16 by 2½ foot bin in the center of the excavation. The bin was insulated and waterproofed, then filled with 20 tons of washed rock. This was the storage for solar heated air from the collector.

Footings and walls were built of pressure treated lumber and ¾ inch thick plywood. Perforated plastic pipe was laid around the perimeter of the footing, for soil water drainage; and the wooden exterior walls were vapor

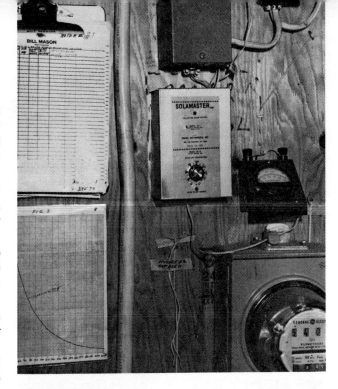

Bill Mason installed control and recording equipment to keep track of how his combined system operates.

Warm air from Franklin type stove is ducted to the return air system of main blower, and distributed throughout the Mason home.

157

proofed with black plastic cemented to the plywood. Walls were framed with pressure treated 2 by 6 lumber; wall cavities were filled with fiberglass insulation. When the walls were up and sealed, earth was back-filled against the house on north, east, and west sides to within about 1½ foot of the roof eaves. The south wall is exposed.

The 400 square foot solar collector is covered with one layer of special low lead, tempered glass and a layer of plastic under the glass. During sunny weather, the collector fan pushes heated air through the 20 tons of rock in the bin below the floor of the 1,800 square foot house and studio. On its way to the storage bin, the heated air passes over and around three 62-gallon preheater tanks for the domestic hot water supply.

Water from a 300 foot deep well comes to the surface at 56-58 degrees F. The hot water line is T'd off to the preheater tanks located in the solar air duct, where the water is heated to about 75-80 degrees. This preheated water then goes to an electric water heater, which adds 40 or 50 degrees to the domestic tap water.

The air distribution blower on the electric furnace distributes all warm air to living spaces, regardless of the source of heat: solar, wood stove, electric heat pump or 20 kw electric furnace. A "Solamaster" controller manages the entire heating system, except for the wood burning stove.

"We installed a large air return grill in the living room near the wood stove," says Bill Mason. "The heated air from the stove is ducted to the return air plenum of the electric furnace. We have a manual override on the furnace blower, so that we can operate it to pull stove heated air through the heating duct to all parts of the house."

Mason's solar heat storage will provide enough heat to keep living spaces at 70 degrees F. for about two days, except in extremely cold, windy weather. Then, a heat pump automatically picks up the load. If outside air is too cold for the machine to extract enough heat for living spaces, the controller shifts to the 20 kw furnace. The furnace actually runs very little of the time, even on coldest days.

Mason designed his earth-contact solar-wood heated home somewhat along the lines of a plan developed by Jerry Newman, a U.S. Department of Agriculture rural housing specialist at Clemson University. During the winter of 1979-80, Newman installed electric meters, strip recorders and other monitoring equipment to keep tabs on Mason's heating system.

"The readings aren't all analyzed yet," says Mason. "But we have found that the entire system works best when we operate the various components to complement each other. Solar heat plus wood heat plus heat pump is often more economical and more comfortable than trying to make the solar system do it all."

The illustrations given are only three examples of hundreds of U.S. homeowners who are fighting their own private skirmishes against a worsening national energy crisis—and winning. There are individuals in every region of the country who are proving, by example, that our energy problems do not have to be fatal. These homeowners are building and operating systems to make the best use of those natural sources of energy that are most abundant in their region.

And, in true "good neighbor" fashion, most of the private energy pioneers are willing to share their experiences. As you plan energy saving and energy producing investments, talk to homeowners who have blazed a trail into this new frontier. They will share their experiences with you—both the ones that work well and those that don't work at all, and you can learn from both.

11
THE MONEY SIDE OF ENERGY

"Money is what things run into, and people run out of."

Mark Beltaire
Detroit Free Press

Sunshine is free. No one owns the wind. But building the pipelines to harvest energy from sun, wind, water, wood, and geothermal sources can run into a pretty penny.

Recent legislation by the U.S. Congress and several state lawmaking bodies provide direct financial assistance for energy investments, generally in the form of income, property and sales tax credits. Several federal and state programs also make low interest, long term loans available for homeowners who add energy saving features to existing houses, or build new alternate energy systems. Government agencies, as well as private businesses, associations, and foundations offer research grants to individuals building "experimental" natural energy systems. This chapter reports on various federal and state financial legislation in effect at this time.

However, more and more tax assistance programs are being put into law all the time, and some programs expire. For example, the U.S. Department of Energy funded a loan program for feasibility studies of small hydroelectric projects—up until November, 1979, when the program ended. So, your best bet is to contact those federal, state, and local government offices mentioned, to find out the current status of programs when you get ready to buy or build. Throughout this chapter, you'll find addresses of agencies concerned with the money side of energy.

FEDERAL TAX CREDITS

Your biggest tax saver on energy investments is Uncle Sam's residential energy tax credit, provided for in the National Energy Act that became law in November, 1978.

Notice that these credits are dollar for dollar reductions in the taxes you owe the Internal Revenue Service, not merely deductions from taxable income.

You can write off up to $2,200 in tax payments on investments to power your home with alternate sources of energy which qualify: solar, wind, or geothermal systems. The credit is 30 percent of the first $2,000 invested and 20 percent of the next $8,000.

The credit is non-refundable, and the equipment must be installed before Jan. 1, 1986. However, excess credit can be carried forward until 1987. In other words, if you install natural energy equipment in 1985 that costs $10,000, you are entitled to the full $2,200 tax credit. But suppose you only report $1,800 taxes due (from your business or withholding statement) on your 1985 federal tax returns. You can subtract $1,800 of the total credit and thus owe *no* federal taxes in 1985, then carry the remaining $400 credit over to the 1986 tax returns.

The rules are somewhat hazy on what's allowed. Most passive and active solar systems for space heating and domestic water heating qualify, as does equipment to tap geothermal sources for space and water heating. Wind powered electrical generating plants also earn the credit.

An attached solar greenhouse may not qualify, however, unless it is built *primarily* to collect solar heat for dwelling space. To be on the safe side, if you build a greenhouse to serve as a solar heat collector for living space or domestic water heating, don't call it a greenhouse on tax forms. The device can be referred to as a "high volume solar energy collector." After all, that's what it is.

You can also claim up to $300 in tax credits for money spent on insulation, weatherstripping, caulking and other energy saving materials, including installing a more

Uncle Sam's tax incentives for energy saving allows up to a total of $2,500 to be deducted directly from income taxes due.

energy-efficient furnace. The credit is 15 percent of the first $2,000 spent, and expires Dec. 31, 1985. New storm doors, storm windows, clock thermostats, and electric ignition systems for furnaces and water heaters may also qualify for this conservation credit.

The White House has proposed a similar 15 percent tax credit for the purchase and installation of wood burning equipment, but as of this writing, Congress is still meditating on that one.

LOANS FOR ENERGY

"Lenders have been reluctant to finance underground homes and some solar installations," says Chip Whittier, Norman, Oklahoma, energy and financial consultant. "There's still some stigma attached by some lenders to alternate energy systems because so-called hippie types were into alternate energy first in many areas."

But that's changing, and changing fast. The Federal Housing Authority of the U.S. Dept. of Housing and Urban Development (HUD) makes guaranteed loans to install solar heating, cooling, and hot water units under Title 1 of the National Housing Act. Funding is a maximum of $15,000, for up to 15 years.

Under the National Energy Conservation Policy Act of 1978, all large public utilities—both electric and natural gas—must offer low cost home energy audits to their customers. Part of the audit report is a list of lenders who might finance the work needed; in some cases the utilities install equipment and finance the cost. A check with your local utility or rural electric cooperative will be worthwhile, if you need some help in finding a lender.

Federal Housing Authority loans help finance natural energy systems, such as these solar homes in Maryland (right) and Virginia. (Courtesy of U.S. Dept. of Housing and Urban Development)

Several public and private agencies make grants for research and demonstration. Bill Mason, shown here, records the performance of his earth-contact, solar heated home for Clemson University engineers. In return, the University pays Mason's electric bills.

The most ambitious energy lending program to date is the proposed Solar Energy Development Bank that would provide low interest (3% to 8%) loans to owners and builders of homes, who buy and install solar energy systems. Loan limits would be $10,000 for single family dwellings, with up to 30 years pay-back period. The projected funding level for the fiscal year ending Aug. 30, 1980, would be $100 million; $150 million for the next year; and $200 million for the year ending Aug. 30, 1982.

Farmers can borrow money from the U.S. Department of Agriculture's Agricultural Stabilization and Conservation Service to construct solar grain drying systems.

GRANTS FOR RESEARCH

Several government agencies, but principally HUD and the U.S. Department of Energy (DOE), have grants-in-aid programs to fund energy developments that qualify as "research" or "demonstration" projects. DOE's Office of Energy Technology accepts proposals for research and development programs on solar, geothermal and other alternate energy technologies.

Individuals, local non-profit organizations and small businesses can also apply for "appropriate technology"

grants for small scale, decentralized projects maximizing the use of renewable energy resources. Grants of up to $10,000 are made for good ideas, and up to $50,000 for the design and assembly of a demonstration model of an approved system.

However, there's one clinker in the federal government grant program. Under federal law, the government takes title to any inventions stemming from grant funded research. In some cases, title to the invention is returned to the inventor through a waiver. If you request grant money from a U.S. government agency, and have any inkling that a patentable invention may result from the work being financed, it's a good idea to include a waiver request in the application for the grant.

The National Center for APPROPRIATE Technology is an independent, non-profit corporation created to develop and apply "appropriate" technologies to specific needs of low income people. The Center is funded by the U.S. Community Services Administration, and issues grants for the development and demonstration of alternate natural energy systems.

During its first two years of operation, the Center awarded nearly $2 million worth of grants to fund 173 projects. Generally, the Center provides funding for projects sponsored by Community Action Agencies and other local, non-profit groups.

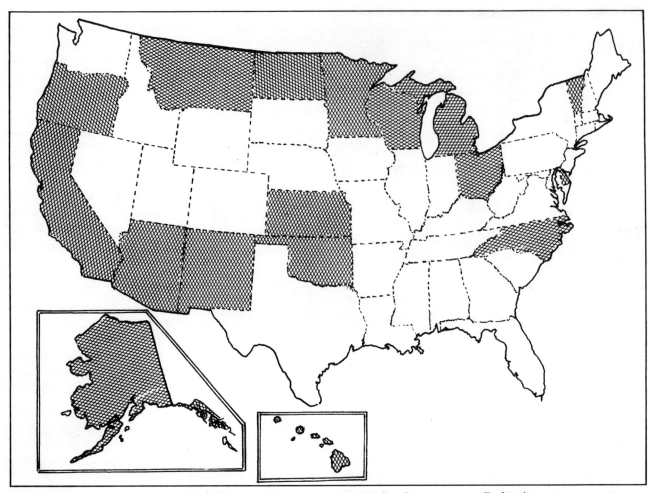

Shaded areas show states which allow income tax credits for alternate energy systems.

STATE ENERGY LEGISLATION

State legislatures have passed a number of laws favoring energy conservation and natural energy development. Here's a summary of what's on the books now. But new laws are being enacted all the time. Contact your state's appropriate agency to bring yourself up to date on taxation, grants, loans and other programs.

Alaska   Allows a 10% residential conservation credit for up to $200 for money spent on insulation, storm windows, solar, wind, tidal, and geothermal energy systems.

Contact State Dept. of Revenue, State Office Bldg., Juneau, AK 99811 907/465-2326.

Arizona

Arizona allows tax credits for solar energy for 30% of the cost, to a maximum of $1,000 credit. Solar energy devices are also exempt from property, transaction and use taxes. A 35% credit is allowed for insulation, ventilating devices, storm windows and doors, up to a maximum of $100. The state also provides authority for local governments to regulate solar access. After June 30, 1980, any active solar system installed must be installed by a licensed contractor.

Contact Dept. of Revenue, Box 29002, Phoenix, AZ 85038 602/255-3381

Arkansas

Individuals may deduct the entire cost of solar heating and cooling equipment, insulation, storm windows, and motor-driven ventilating devices from taxable income. Some biogas systems, geothermal, hydroelectric, wind energy, and wood-burning devices other than fireplaces also qualify.

Contact Dept. of Revenue, 7th and Wolfe, Little Rock, AR 72201 501/371-2193.

California

Personal income tax credits of 55% of the cost of solar energy systems, up to a maximum of $3,000.

Contact Franchise Tax Board, Sacramento, CA 95807 916/355-0370.

California's Solar Energy Demonstration Loan program provides $2,000 in interest free loans for solar space heating and domestic hot water systems in some areas of the state.

Contact Dept. of Housing and Community Development, 921 Tenth St., Sacramento, CA 95814 916/455-4728.

Veterans applying for home loans may borrow an additional $5,000 if the home is equipped with solar heating devices.

Contact Dept. of Veteran Affairs, P.O. Box 1559, Sacramento, CA 95807.

California is one of few states to enact "solar rights" laws. Anyone who owns or controls real estate is prohibited from allowing a tree or shrub to cast a shadow on a solar collector between 9:30 AM and 2:30 PM.

Colorado

Solar heating and cooling devices are assessed for property tax at 5% of value. Individual taxpayers may deduct the entire cost of solar, wind, and geothermal equipment from taxable income. Colorado recognizes solar easement.

Contact Dept. of Revenue, 1375 Sherman St., Denver, CO 80261 303/839-3781.

Connecticut

Solar collectors are exempt from sales tax through Oct. 1, 1982. Local governments may exempt solar heating, cooling and electrical equipment; windmills; and water wheels from property taxes.

Contact State Tax Dept., 92 Farmington Ave., Hartford, CT 06115 203/566-2601.

Connecticut also has a low interest loan fund for insulation and alternate energy equipment, up to $3,000.

Contact Commissioner of Economic Development, 210 Washington St., Hartford, CT 06115.

Delaware

Income tax credits up to $200 for solar hot water systems.

Contact Div. of Revenue, 820 French St., Wilmington, DE 19801 302/571-3360.

Florida

Solar energy systems are exempt from sales tax until June 30, 1984.

Contact Dept. of Revenue, Carlton Bldg., Tallahassee, FL 32304 904/488-6800.

Georgia

Taxpayers may claim refund of sales tax paid on solar equipment.

Contact Dept. of Revenue, 309 Trinity-Washington Bldg., Atlanta, GA 30334 404/656-4065.

Hawaii

A 10% income tax credit for any non-nuclear, non-fossil fuel system installed before Dec. 31, 1981. Also, property tax exemptions for solar energy systems.

Contact State Tax Dept., P.O. Box 259, Honolulu, HI 96809 808/548-3270.

Idaho

Solar, wood, wind and geothermal systems qualify for income tax deductions up to $5,000 in any one year, at the rate of 40% of cost in the first year and 20% of the cost in each of the next three years. Fireplaces with controlled draft and heat circulating blowers qualify.

Contact State Tax Commission, 5257 Fairview, Boise, ID 83772   208/384-3290.

Illinois

Contact local assessor for details on an alternate property tax valuation for alternate energy systems. Illinois also has funded a $5 million solar energy demonstration program.

Contact Dept. of Energy, 222 South College, Springfield, IL 62706   217/782-7500.

Indiana

Contact local assessors for special property tax valuation for dwellings with alternate energy systems.

Iowa

State law requires that solar energy systems will not increase assessed valuation through 1985. Iowa also has a loan and grant fund for property improvements by low income families; solar systems qualify.

Contact Housing Finance Authority, 218 Liberty Bldg., Des Moines, IA 50319   515/281-4058.

Kansas

Income tax credit of 25% of cost of the solar and wind energy systems, to a maximum of $1,000. If a solar energy system supplies 70% or more of the energy for heating and cooling, state refunds up to 35% of property tax for five consecutive years, through 1985.

Contact Dept. of Revenue, P.O. Box 692, Topeka, KS 66601 913/296-3909.

Louisiana

Solar energy equipment installed in owner-occupied dwellings and swimming pools exempted from property tax.

Contact local parish assessors.

Maine

Solar space and water heating systems are exempt from property tax for five years after installation. Purchasers of solar energy systems may apply for sales tax refunds. Maine protects access to direct sunlight for solar energy users.

Contact local planning and zoning groups for solar access; for sales tax rebate, Office of Energy Resources, 55 Capitol St., Augusta, ME 04330   207/289-2196

Maryland

Solar energy systems are assessed at no more than a conventional system needed to serve the building. City and county governments may offer property tax credits for solar equipment.

Contact local assessors.

Massachusetts

Solar energy systems are exempt from property tax for 20 years from date of installation. Sales of solar, wind and heat pump equipment are exempt from sales tax.

Contact Dept. of Corporation and Taxation, 100 Cambridge St., Boston, MA 02204.

Massachusetts state laws also authorize banks and credit unions to make long term loans for solar installations, up to $15,000 and $12,000, respectively.

Michigan

Equipment used for solar, wind or water energy is exempt from state excise tax and property taxes. Income tax credits can be claimed for solar, wind, and water energy devices, at the rate of 25% for the first $2,000 spent, plus 15% of the next $8,000. Swimming pool heaters qualify if 25% or more of their heating capacity is used for residential heating. Some wood furnaces also qualify.

Contact State Tax Commission, State Capitol Bldg., Lansing, MI 48922   517/373-2910.

Minnesota

Individual income tax credits of 20% of the first $10,000 spent on renewable energy equipment, including passive and active solar, wind, geothermal, earth-contact and underground dwellings, and equipment to produce ethanol, methanol and methane gas for fuel. Also, these systems are exempt from property tax. All installations must be made prior to Jan. 1, 1984.

Contact Dept. of Revenue, 658 Cedar St., St. Paul, MN 55145 612/296-3781

Mississippi

Some sales tax exemptions for labor and equipment for solar heating, lighting and electricity generation.

Contact State Tax Commission, P.O. Box 960, Jackson, MS 39205   601/354-6274.

Missouri

Recognizes solar energy use as a property right, and recognizes solar easements. Some loan and grant money for home weatherization.

Contact Dept. of Natural Resources, P.O. Box 1309, Jefferson City, MO 65102.

Montana

Income tax credits of 10% for first $1,000 and 5% of next $3,000 spent on active and passive solar, wind, solid waste, biogas, wood and small scale hydroelectric equipment. Also provides deductions for energy conservation costs, including insulation and storm windows.

Contact Dept. of Revenue, Mitchell Bldg., Helena, MT 59601 406/449-2837.

Montana law also permits utility companies to install and finance energy conservation materials and non-fossil fuel energy systems at two percentage points below the interest rate in the Ninth Federal Reserve District.

Contact local utility.

Nebraska

Recognizes solar easements.

Nevada

Property tax allowances for solar, wind, geothermal, water, and solid waste energy systems.

Contact local assessors.

The Nevada Dept. of Energy establishes energy conservation standards, as well as design and construction standards for solar, geothermal, wind and other renewable energy systems, to be included in all city and county building codes.

Contact Dept. of Energy, 1050 E. Williams, Carson City, NV 89710.

New Hampshire

Cities and towns grant property tax exemptions for solar energy systems and some wood fired central heating systems.

Contact local assessors.

New Jersey

Solar heating and cooling, sea thermal gradient systems and wind powered systems are exempt from state sales tax.

Contact Div. of Taxation, P.O. Box 999, Trenton, NJ 08646 609/292-6400.

New Mexico

Income tax credit of 25% of the cost of solar energy equipment, up to $1,000, including swimming pool heaters. Also allows income tax credits for some solar irrigation systems. Credit in excess of taxes due is refunded.

Contact Dept. of Taxation and Revenue, P.O. Box 630, Santa Fe, NM 87503   505/827-3221.

New York Property tax reductions for solar and wind energy installations made before July 1, 1988.
Contact local assessors.

North Carolina Income tax credit of 25% of the cost of solar heating, cooling, and hot water systems, up to $1,000 maximum.
Contact Dept. of Revenue, P.O. Box 25000, Raleigh, NC 27640 919/733-3991.

North Dakota Income tax credits of 5% per year for two years on solar and wind energy devices. Solar heating and cooling systems are exempt from property tax for five years after installation.
Contact State Tax Commission, Capitol Bldg., Bismarck, ND 58505 701/224-3450.

Ohio Solar, wind and hydrothermal energy systems installed through Dec. 31, 1985, are exempt from real estate taxes. Income credit of 10% of the cost of these systems may be taken to a maximum of $1,000.
Contact Ohio Tax Commission, 1030 Freeway Drive, Columbus, OH 43229 614/466-7910.

Oklahoma Income tax credits for solar energy devices of 25% of the cost, to a maximum of $2,000. The credit may be taken only once, but the total amount can be applied on taxes for up to three years.
Contact State Tax Commission, 2501 Lincoln Boulevard, Oklahoma City, OK 73194 405/521-3125.

Oregon Laws provide for income tax credits to 25% of the cost of solar, wind, or geothermal energy systems, to a maximum credit of $1,000. In addition, solar, geothermal, wind, water, and methane gas systems are exempt from property tax until Jan. 1, 1998.

Contact Dept. of Revenue, State Office Bldg., Salem, OR 97310 503/378-3366.

Loan funds for alternate energy systems are available through Dept. of Energy, Labor and Industries, Salem, OR 97310 503/378-4128.

Oregon veterans may borrow in excess of the maximum home loan limit for solar energy systems.
Contact Dept. of Veterans Affairs, 3000 Market Street Plaza, Salem, OR 97310 503/378-6438.

Oregon laws also limit the liability of landowners who permit others to cut firewood on their property.

Rhode Island Special assessment rates for residences equipped with solar heating and cooling systems.
Contact local assessors.

South Carolina Has adopted statewide building code standards with energy efficiency criteria.
Contact Div. of Energy Resources, 1122 Lady St., Columbia, SC 20201.

South Dakota Property tax assessments credit for solar, wind, geothermal and biomass energy systems.
Contact local assessors.

Tennessee Solar and wind energy systems are exempt from property taxes.
See local assessors.

State provides loans to low and medium income persons to make energy-saving improvements, including solar hot water systems.
Contact Housing Development Authority, Hamilton Bank Bldg., Nashville, TN 37219 615/741-3023.

Texas

Solar and wind powered devices are exempt from property taxes. Solar energy systems for heating, cooling or electrical power are also exempt from state sales tax.

Contact Comptroller of Public Accounts, Capitol Station, Austin, TX 78775.

Utah

Recognizes solar easements.

Vermont

Solar, wind, and wood fired central heating devices are eligible for income tax credit of 25% of the cost, up to $1,000 maximum credit. Also, many towns exempt alternate energy equipment from property tax.

Contact State Tax Dept., State St., Montpelier, VT 05602 802/828-2517.

Virginia

Any county, town or city may exempt solar devices from property tax, if the equipment is certified by the State Board of Housing.

Contact local tax governing body.

Virginians now are voting on a constitutional amendment referendum to enact a statewide property tax exemption for alternate energy systems.

Washington

Solar space and water heating systems have a seven-year exemption from property taxes, if claims are filed before Dec. 31, 1981.

Contact local assessors or boards of adjustment.

Wisconsin

A direct subsidy program refunds a portion of the cost of alternate energy systems. Rate of refund is 24% in 1979 and 1980; 18% in 1981 and 1982; and 12% in 1983 and 1984. Subsidy is paid on up to $10,000 cost.

Contact Dept. of Industry, Labor and Human Relations, 201 E. Washington, Madison, WI 53702 608/266-1149.

Additional Information

For updated information on federal legislation and programs affecting solar energy, write to Solar Heating Center, P.O. Box 1607, Rockville, MD 20850. The Center operates a toll-free telephone, 800/523-2929 (In Pennsylvania, 800/462-4983).

DOE's Office of Energy Technology accepts proposals for research and development grants from individuals and organizations. Write to Unsolicited Proposal Management, Division of Procurement, U.S. Department of Energy, Washington, D.C. 20545. DOE also provides information on all energy sources and energy conservation technologies. Contact the DOE Technical Information Center, P.O. Box 62, Oak Ridge, TN 37830.

To learn more about the programs sponsored by the National Center for Appropriate Technology, contact the Center at P.O. Box 3838, Butte, MT 59701 (Telephone: 406/494-4572).

For more information on government guaranteed home improvement loans for energy devices, contact the Federal Housing Authority, U.S. Department of Housing and Urban Development, Washington, DC (Telephone: 202/755-5284).

Home buyers are more and more conscious of energy saving features, such as this solar assisted heat pump installation in Hazelton, Pennsylvania. (Courtesy of Grumman Energy Systems)

If you're planning to buy, install, or build a natural energy system anytime soon, start checking with federal, state, and local agencies early on loans, grants and tax incentives that can put money in your pocket. The government is finally giving more than mere "lip service" to natural energy systems. Also inquire into community and private organizations that may help fund alternate energy systems.

HOW ABOUT RESALE VALUE?

The demand for housing in general depends on national and local economic conditions, as well as some "style" features that may currently be in vogue. For example, a few years back, the most popular type of home was the sprawling, California type ranch home.

Today, energy saving features are coming into hot demand. This includes both energy conserving design features, and alternate energy systems incorporated. However, not every potential house buyer is quite ready to accept underground dwellings, or solar devices that make a home's appearance conspicuously different from others in the neighborhood.

But the swing to homes designed and equipped to save energy is picking up momentum. The cost to heat and power any home is a major buying consideration. With each jump in the cost of petroleum fuels, more potential home buyers move energy efficiency higher on the list of important features for which to look.

In most regions of the U.S., a home equipped with proven workable energy saving, energy producing systems will be worth enough extra dollars to pay for the cost of installing the devices. Well built alternate natural energy systems have an expected life of 20 years or more, and many buyers are willing to trade future savings on energy for a bigger initial investment.

12
COMMUNITY ENERGY GROUPS

"Too often, citizen volunteerism is written off as futile altruism, when in fact the benefits which citizens produce by their involvement are very real, both materially and spiritually. That gives me hope."

Donn Werling
Epoch B Project
Evanston, Illinois

The era following the Second World War saw Americans spread across the land adjacent to big cities. Suburbs and subdivisions grew up with little concern for mass transportation availability: Everyone owned a couple of cars and gasoline was cheap.

Houses were built in "bedroom" communities located miles from where the occupants worked. And, they were built with little regard to their efficient energy consumption: electricity and heating fuel were cheap and plentiful.

But few of these suburbs and subdivisions have grown into *communities* in the best sense of the word. The term "bedroom" is an apt adjective for many of these developments. Residents work, play and shop outside the area, and return there primarily to sleep and change clothes.

Today, in many areas of the U.S., the energy crisis is giving back to Americans a long lost sense of "neighborhood" not in much evidence since World War II. We have a common enemy—fuel scarcity—to fight and beat. More and more local residents are losing faith in big, centralized organizations to provide heat and power. In several communities, people are huddling together to warm themselves with ideas and local self-help programs. And they are finding that citizens can do something on a local, neighborhood level if they have the drive and will.

"As an environmental educator, I feel that the process outlined (community action) exemplifies not only the best, but the only viable strategy currently available for a community to sincerely address itself to the energy dilemma in a cost-effective way that not only stays true to, but

strengthens our democratic traditions," says Donn Werling, environmentalist who directs the Evanston, Ill., Ecology Center.

Werling and other Evanston residents have spearheaded among the more successful community projects in alternate energy, called "Epoch B." We'll go into more detail shortly on how a group of concerned Evanston citizens got Epoch B underway.

However, this is not the only community action program in the country working on energy problems. From the heart of New York City to California, citizens are forming their own energy action groups.

Take the case of Davis, California. Concerned citizens, builders, architects, city fathers, students and faculty from the nearby University of California took energy matters into their own hands in October, 1975. City building codes were rewritten from the prescriptive language that specifies materials and details of construction, to standards that specify *performance* and leave the selection of materials and techniques up to builders. So long as the finished structure meets rather stringent heat loss standards, builders have pretty much a free hand in deciding how to meet those standards.

One result of the community efforts of Davis is that energy consumption in the city has dropped by nearly 20 percent since the program began. A locally funded, locally planned solar housing development, called "Village Homes," is getting underway there.

You might not expect to find New York's Lower East Side the site of a dynamic energy independence effort.

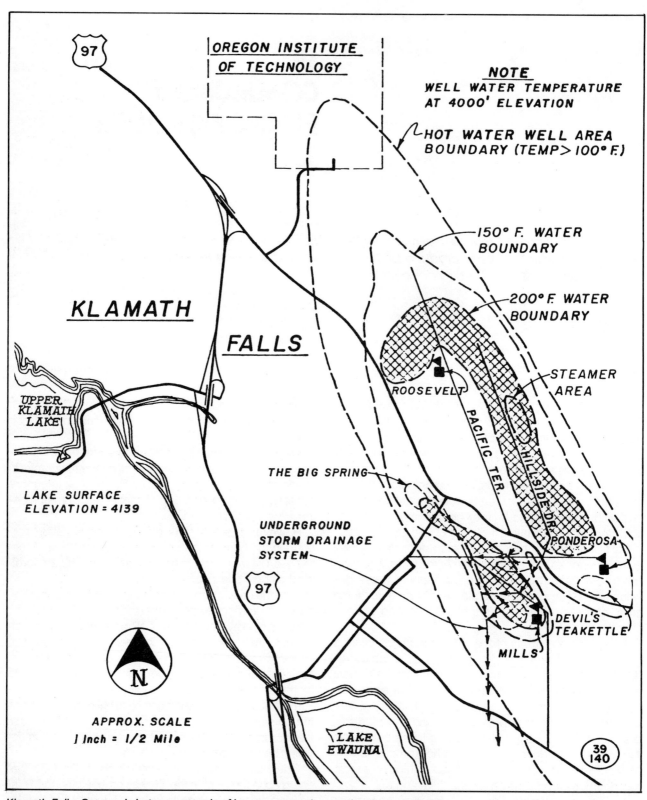

Klamath Falls, Oregon, is but one example of how a community can develop a regional natural energy resource—in this case, geothermal energy, pumped from the cross-hatched area on the map. (Courtesy of Oregon Institute of Technology)

Across the country, concerned citizens are spurring their communities in local action energy programs.

That particular area of the "Big Apple" has been more often characterized by stripped autos and "fire-for-hire" building burn-outs. But a community "Energy Task Force" provides technical and educational assistance to low-income housing groups working to improve and revive neighborhoods.

One of the more conspicuous successes of the Energy Task Force is a building on East 11th Street. With some funding from the Community Services Administration, the tenant building was insulated, weatherized, equipped with solar heating and a wind powered generator that is tied to Consolidated Edison's power grid with a synchronous inverter. The group now is working in other low income areas of the city.

The Epoch B project at Evanston, Illinois, mentioned earlier, is an education-demonstration program on insulation, weatherization, solar and wind energy that grew out of the concerns of a handful of people. Evanston is a city of 80,000 or more, perched on the shore of Lake Michigan just north of Chicago. It's the home of Northwestern University and national headquarters for a number of organizations.

The original committee that formed Epoch B consisted of two housewives, the director of the Evanston Ecology Center, an associate professor of physics, and an elementary school teacher. As the project grew, city leaders, engineers, architects, environmentalists, and faculty members from Northwestern became interested.

"In October, 1976, Epoch B first went public with hopes and dreams for the future," says Libby Hill, currently director of the project. "We knew there was a need for an education-working demonstration program on energy,

but we certainly never predicted then how quickly the energy question would accelerate into a *crisis.* Of the fifty or so people at the first session, we had skeptics who wondered about our sanity to have plans and no money. But a majority were caught up in the excitement and promise of the project and they have given it life."

In the best traditions of "show and tell", Epoch B set out to make an educational demonstration, on a typical northern urban building, of particular ways to use energy wisely. They had such a structure readily at hand: the Evanston Ecology Center building. But, before they could show and tell, Epoch B members had to *learn.*

"We started researching energy conservation and alternate energy subjects," says Ms. Hill. "We were interested and impressed with the scope of solar projects throughout the country, but found that most of them were produced by industrial or large govenment funded programs, were research projects by universities, or were communes and basement tinkerers. Our spontaneous community project didn't seem to fit any of these categories."

After boning up on their subject matter, Epoch B members proposed to the city of Evanston that the project use the Ecology Center as a physical facility to demonstrate how an existing building in an urban area could be adapted for environmentally sane, appropriate energy technology. From the first, members of Epoch B were determined that the project would belong to those who worked on it, and would produce practical "take home" information on energy conservation and alternate energy systems.

"We decided to try in every way possible to do what

was most basic, what was clearly essential, by the most simple methods," says Ms. Hill. "We discovered that the quality of the program was assured as long as we kept strictly to this idea."

The people who were active in Epoch B didn't merely want to talk and read about ways to solve energy problems; they wanted to *do.* By this time, the project had attracted several people with experience, either as professionals or amateurs, in various aspects of building and engineering. So, using the Ecology Center building as a working model, Epoch B set up a series of "learn by doing" workshops.

"We designed each workshop to be valuable in and of itself, so that it would fit together into an integrated whole and follow a logical progression," says Ms. Hill.

Here are the workshop steps:

1. Learn where heat is being lost from the building, and how to calculate how much heat is being lost.

2. Learn about appropriate insulation possibilities and how to make the building thermally efficient.

3. Learn how to construct and install solar collectors.

4. Learn how to construct and install an active solar heating system and heat storage.

5. Learn about the electrical demands of the building; how much wind power is available; then design and install a wind powered generator.

By December, 1979, Epoch B had completed all its initial building programs. A new solar greenhouse was being finished and the "shake-down" operation of a Cycloturbine wind energy machine with an induction generator was getting underway.

The Epoch B group now is exporting the knowledge and experience gained to others; through high school summer classes, booklets, slide shows and on-the-spot technical help. The group also conducts a "Project Conserve" home energy audit program for Evanston homeowners.

Perhaps the biggest feather in the Epoch B cap is a new recreation center that will have part of its energy needs met by a passive solar system. The Evanston Recreation Board had been lukewarm to the idea of using solar energy in public buildings, but the Evanston City Council voted to build the center.

"As we approach the end of the first major building program of Epoch B, we think it again wise to meet to see where we should go from here," says Libby Hill.

Whatever new directions the community action group takes, it will be a case of involved local people coming to grips with an energy problem that affects their lives. Be it Davis, California; East 11th Street in New York City; or Evanston, Illinois; groups of citizens are proving that they can take some local control of their lives, and work together to meet and solve problems.

As painful as ever-scarcer, ever-higher fuel may be for American homeowners, to the extent that our common problems with energy make us work together to solve those difficulties, maybe the fuel crisis ain't necessarily fatal after all.

In Appendix II, you will find addresses to write for more information on Epoch B and other community scale action programs developed to let local people take a hand in solving energy related problems.

APPENDIX I
Metric Conversion
Tables

The United States is steadily going to the SI metric system of measurement. Some manufacturers have already shifted to metric measurements (such as liquor bottlers, who have forsaken pints, fifths and quarts for liters) and more will as time goes along.

Instead of feet, quarts, pounds and miles, we'll be using meters, liters, grams and kilometers. The metric system is a standardized index based on 10. There are 10 millimeters to a centimeter, 10 centimeters to a decimeter, 10 decimeters to a meter, and so on.

Here are conversions of U.S. to Metric, and Metric to U.S. measure for length, square area, volume or capacity, mass or weight and temperatures:

Length

U.S. to Metric
1 inch = 25.40 millimeters
1 inch = 2.540 centimeters
1 foot = 30.480 centimeters
1 foot = 0.3048 meter
1 yard = 91.440 centimeters
1 yard = 0.9144 meter
1 mile = 1.609 kilometers

Metric to U.S.
1 millimeter = 0.03937 inch
1 centimeter = 0.3937 inch
1 meter = 39.37 inches
1 meter = 3.2808 feet
1 meter = 1.0936 yards
1 kilometer = 0.62137 mile

Area

U.S. to Metric
1 sq. inch = 645.16 sq. millimeters
1 sq. inch = 6.4516 sq. centimeters
1 sq. foot = 929.03 sq. centimeters
1 sq. foot = 0.0929 sq. meter
1 sq. yard = 0.836 sq. meter
1 acre = 0.4047 sq. hectometer
1 acre = 0.4047 hectare
1 sq. mile = 2.59 sq. kilometers

Metric to U.S.
1 sq. millimeter = 0.00155 sq. inch
1 sq. centimeter = 0.1550 sq. inch
1 sq. meter = 10.7640 sq. feet
1 sq. meter = 1.196 sq. yards
1 sq. hectometer = 2.471 acres
1 hectare = 2.471 acres
1 sq. kilometer = 0.386 sq. mile

Volume or Capacity

U.S. to Metric
1 fluid ounce = 2.957 centiliters = 29.57 cubic centimeters
1 pint = 4.732 deciliters = 473.2 cubic centimeters
1 quart = 0.9463 liter = 0.9463 cubic decimeter
1 gallon = 3.7853 liters = 3.7853 cubic decimeters

Metric to U.S.
1 centiliter = 10 cubic centimeters = 0.338 fluid ounce
1 deciliter = 100 cubic centimeters = 0.0528 pint
1 liter = 1 cubic decimeter = 1.0567 quarts
1 liter = 1 cubic decimeter = 0.26417 gallon

173

Mass or Weight

U.S. to Metric
1 ounce (dry) = 28.35 grams
1 pound = 0.4536 kilogram
1 short ton (2,000 lbs.) =
    907.2 kilograms
1 short ton = 0.9072 metric tons

Metric to U.S.
1 gram = 0.03527 ounce
1 kilogram = 2.2046 pounds
1 metric ton = 2.204.6 pounds

1 metric ton = 1.102 short tons

Celsius	Fahrenheit	
100	212	—Water boils at sea level.
98	208	
96	205	
94	201	
92	198	
90	194	
88	190	
86	187	
84	184	
82	180	
80	176	
78	172	
76	169	
74	165	
72	162	
70	158	
68	154	
66	151	
64	147	
62	144	
60	140	
58	136	
56	133	
54	129	
52	126	
50	122	Range for domestic hot water
48	118	
46	115	
44	111	
42	108	
40	104	
38	100	
36	97	
34	93	
32	90	
30	86	
28	82	
26	79	
24	75	
22	72	
20	68	
18	64	
16	61	
14	57	
12	54	
10	50	
8	46	
6	43	
4	39	
2	36	
0	32	—Water freezes at sea level

Degrees of Temperature

To convert degrees of Fahrenheit to Celsius, subtract 32° and multiply by 5/9ths. For example, 77 degrees F. would be 25 degrees C. (77°F. - 32 = 45, and 5/9ths of 45 is 25°C.). The process is reversed to convert Celsius to Fahrenheit. Multiply the Celsius temperature by 9/5ths and *add* 32 (25°C. x 9/5 = 45 plus 32 = 77°F.).

APPENDIX II
For More
Information...

ENERGY CONSERVATION

Energy Savers Catalog, by editors of *Consumers Guide* (G.P. Putnam and Sons, New York, NY). A guide to energy efficient mechanical and electrical appliances and equipment.

How to Cut Your Energy Bills, by Ronald Derven and Carol Nichols (Structures Publishing Co., Farmington, MI, 1976). A complete guide to conserving energy by insulating, weather stripping and general "tightening up."

In the Bank.. Or Up the Chimney? (U.S. Dept. of Housing and Urban Development, Washington, DC) A dollars and cents guide to energy saving home improvements.

One Hundred Ways to Save Energy and Money in the Home (Office of Energy Conservation, Ottawa, Ontario, Canada, 1975). The title describes the book well.

Saving Home Energy, by Richard V. Nunn (Oxmoor House, Birmingham, AL, 1975) Guide to insulating and weatherstripping existing homes.

The Energy Efficient Home, by Steven Robinson and Fred S. Dubin (New American Library, New York, NY, 1978). An excellent guide to energy conservation, with sections on solar, wood, and wind energy.

Tips for Energy Savers (U.S. Government Printing Office, Washington, DC). Booklet of energy conserving tips for homeowners.

Your Energy-Efficient Home, by Floyd Hickok (Prentice-Hall, Inc., Englewood Cliffs, NJ). A somewhat technical book on how to engineer your home for greater energy efficiency.

ALTERNATE ENERGY — GENERAL

Alternate Sources of Energy, Milaca, MN 56353. Bimonthly magazine that covers the alternate energy scene thoroughly; special theme issues on solar, wood, wind, underground housing, etc.

Energy for Rural Development (National Academy of Sciences, Washington, DC, 1976). Report of an Ad Hoc committee on renewable resources and alternate technologies for developing countries. Useful projected timetable for some newer technologies.

Handbook of Homemade Power, by editors of *Mother Earth News (The Mother Earth News,* Hendersonville, NC, 1974). A useful account of several "backyard" type alternate energy devices to harness heat and power from the sun, wind, wood, water, and methane gas.

Producing Your Own Power, edited by Carol Hupping Stoner (Rodale Press, Emmaus, PA, 1974). Good, practical text on alternate energy and how to apply it. Sections on solar, wood, wind, water, and methane gas. Useful section on how to calculate energy requirements.

Volunteers in Technical Assistance, 3706 Rhode Island Ave., Mt. Rainier, MD 20822. VITA publishes and distributes a variety of publications on all aspects of alternate energy.

ENERGY EFFICIENT DESIGN

Alternative Natural Energy Sources in Building Design, by Albert J. Davis and Robert P. Schubert (Van Nostrand Reinhold Company, New York, NY, 1977). Sourcebook examines energy alternatives available to builders.

Architecture and Energy, by Richard G. Stein (Anchor Press, New York, NY, 1977). Guide to energy conservation in building. Aimed at technical audience, but of interest to general readers.

Building for Self-Sufficiency, by R. Clarke (Universe Books, New York, NY). Guide to combining alternate energy features in building systems.

Design With Climate, by Victor G. Olgyay (Princeton University Press, Princeton, NJ, 1963). Technical but comprehensive analysis of building to suit environmental conditions.

Environmental Design Primer, by Tom Bender (Schocken Books, New York, NY, 1977). Focuses on spiritual and aesthetic topics related to building design.

Homegrown Sundwellings, by Peter Van Dresser (The Lightning Tree, Santa Fe, NM, 1977). Report of low cost, owner built homes featuring solar energy.

Low-Cost, Energy-Efficient Shelter, by Eugene Eccli (Rodale Press, Emmaus, PA, 1976). Handbook for owner builders in the planning stages of construction.

30 Energy-Efficient Houses, by Alex Wade and Neal Ewenstein (Rodale Press, Emmaus, PA). Good guide for potential owner builder on designing and planning a home.

National Solar Heating and Cooling Information Center, P.O. Box 1607, Rockville, MD 20850. A government sponsored clearinghouse of information on all aspects of solar energy.

Solar Age Magazine, Harrisville, NH 03450. Monthly publication devoted mostly to solar energy, with some articles on other renewable sources of energy.

Solar Energy: Fundamentals in Building Design, by Bruce Anderson (McGraw-Hill, New York, NY, 1977). A textbook for designing solar energy systems.

Solar Greenhouse Guide (Ozark Institute, Eureka Springs, AR). Handbook on designing, building and operating solar greenhouses.

The Solar Greenhouse Book, by James C. McCullagh (Rodale Press, Emmaus, PA). Comprehensive guide to solar greenhouses.

The Solar Decision Book: A guide for Heating Your Home With Solar Energy, by Richard Montgomery and Jim Budnick (John Wiley and Sons, New York, NY). Text emphasizes the design and installation of durable solar energy systems.

The Solar Home Book, by Bruce Anderson (Brick House Publishing Co., Harrisville, NH, 1976). Excellent manual on solar home designs.

SOLAR ENERGY

Building the Solar Home, by Dubin-Bloome Associates (U.S. Government Printing Office, Washington, DC). Booklet with useful design details.

Buying Solar (U.S. Government Printing Office, Washington, DC). Booklet provides guidelines for consumers buying solar energy equipment and components.

DOE Facilities Solar Design Handbook, (U.S. Dept. of Energy, 1978). Guidelines for solar installations in DOE buildings. Informative, but somewhat technical.

Homeowner's Guide to Solar Heating and Cooling, by W. M. Foster (Tab Books, Blue Ridge Summit, PA, 1976). Guide to buying, installing, and maintaining solar heating, cooling, and hot water systems.

How to Build a Solar Heater, by Ted Lucas (Ward Ritchie Press, Pasadena, CA). Elementary sort of treatment of solar hardware.

WIND POWER

Guide to Commercially Available Wind Machines (U.S. Government Printing Office, Washington, DC, 1978). A list of commercial wind plants for pumping water and generating electricity.

Harnessing the Wind for Home Energy, by Dermot McGuigan (Garden Way Publishing, Charlotte, VT, 1978). An overview of wind as a source of household energy.

Power From the Wind, by Palmer C. Putnam (Van Nostrand Co., New York, NY, 1948). Contains a good section on theory of wind power, wind data analysis. Book is now out of print.

Wind Power Digest, 54468 C.R. 31, Bristol, IN. Quarterly publication is a current, comprehensive source of information on wind energy.

Wind Power for Farms, Homes and Small Industry (U.S. Government Printing Office, Washington, DC, 1977). A comprehensive overview of the use of small wind machines.

Wind Power for Your Home, by George Sullivan (Cornerstone Library, New York, NY, 1978). A good reference on how to evaluate wind resources, with plans and instructions for a do-it-yourself plant.

WATER POWER

Hydroelectric Engineering Practices, by J. Guthrie Brown (Gordon and Breach, New York, NY). Thorough, but technical treatment of hydroelectric engineering.

Hydro Electric Handbook, by W. P. Creager and J. D. Justin (John Wiley and Sons, New York, NY). A comprehensive reference on water generated electricity.

GEOTHERMAL AND EARTH-CONTACT HOUSING

Geothermal Energy Utilization, by Edward F. Wahl (John Wiley and Sons, New York, NY). College text on physical and chemical properties of geothermal water and brines.

Earth-Sheltered Housing Designs (The Underground Space Center, Dept. of Civil and Mechanical Engineering, University of Minnesota, Minneapolis, MN, 1978). A valuable guide to siting and planning earth-contact and underground dwellings.

Multipurpose Uses of Geothermal Energy (Oregon Institute of Technology, Klamath Falls, OR). Proceedings of the International Conference on Geothermal Energy for Industrial, Agricultural and Residential Uses.

Underground Designs, by Malcolm Wells, Cherry Hill, NJ 08002. Wells, an architect, features designs for underground structures in this 87-page book.

WOOD HEAT

Heating With Wood, by Larry Gay (Garden Way Publishing, Charlotte, VT). Good general guide on using wood energy.

The Woodburner's Encyclopedia, by Jay Shelton and A. B. Shapiro (Vermont Crossroads Press, 1976). Helpful guide to selecting wood burning equipment.

Woodburning Quarterly, South Minneapolis, MN 55420. Quarterly publication on wood heat.

Wood Heat, by John Vivian (Rodale Press, Emmaus, PA). A thorough-going manual on heating and cooking with wood.

Wood Heat Safety, (Garden Way Publishing, Charlotte, VT). A comprehensive text on causes, cures and preventions of fires and fire hazards with wood burning devices.

CLIMATE AND WEATHER

Climatic Atlas of the United States (Environmental Data Service, National Climatic Center, Federal Building, Asheville, NC 28801). Reports average winds, temperatures, strongest winds, wind direction and other weather data for selected points in the U.S. Atlas and monthly reports are available for each state.

COMMUNITY ACTION

A Community Project in Alternate Energy (Epoch B, Evanston Environmental Assn., Evanston, IL). Report of the education/demonstration program run by citizens of Evanston, Illinois.

County Energy Plan Guidebook, by Alan Okagaki (Institute for Ecological Policies, Fairfax, VA). A step-by-step manual for creating a renewable energy plan at the county level.

Village Home's Solar House Designs (Rodale Press, Emmaus, PA, 1979). Design details of the homes built in the Village Homes subdivision of Davis, California.

Windmill Power for City People (U.S. Government Printing Office, Washington, DC). A report by the Energy Task Force of New York City. Relates the philosophy, problems and successes of the first urban wind energy program.

APPENDIX III
Manufacturers of Natural Energy Equipment

The following is by no means a complete list of U.S. firms engaged in producing natural energy equipment and materials. It is offered merely to give readers an idea of what's available and where; and provides a place to start looking.

Abraham Lincoln once said: "When you ask from a stranger that which is of interest only to yourself, always enclose a stamp." The companies listed below gladly provide potential customers with information on their products. But these firms are not in the business of supplying free information. When you contact a company, it's good manners to enclose a dollar or two to cover the costs of handling and mailing the information you request.

ENERGY CONSERVING MATERIAL

Appropriate Technology Corp.
339 Green St.
Brattleboro, VT 05301
Insulating window shades

Celanese Fibers
1211 Avenue of the Americas
New York, NY 10036
"Polar-Gard" insulation

Insta-Foam Products
2050 N. Broadway
Joliet, IL 60434
Polyurethane foam insulation

InsulShutter, Inc.
110 N. 7th St.
Silt, CO 81652
Insulating window shutters

Owens-Corning Fiberglas Corp.
Fiberglas Tower
Toledo, OH 43659
Fiberglas insulation

Schlegel Corp.
P.O. Box 197
Rochester, NY 14601
Weatherstripping materials

Shutters, Inc.
110 E. 5th St.
Hastings, MN 55033
Insulating window shutters

Thermo Technology Corp.
Box 130
Snowmass, CO 81654
Insulating curtains

WOOD BURNING EQUIPMENT

All-Nighter Stove Works
80 Commerce St.
Glastonbury, CT 06033
Wood stoves

Ashley Wood Heaters
1604 17th Ave.,S.W.
Sheffield, AL 35660
Circulating heaters; furnaces

Cawley Stove Co.
27 N. Washington St.
Boyertown, PA 19512
Air-tight cast iron stoves

Charmaster Products, Inc.
2301 Highway No. 2
Grand Rapids, MN 55744
Dual fuel fireplace/furnaces

Ashley Circulating heater

King Products Division
P.O. Box 128
Florence, AL 35630
 Wood and coal circulators

Longwood Furnace Corp.
Gallatin, MO. 64640
 Dual fuel furnaces

The Majestic Co.
Huntington, IN 46750
 Heat circulating fireplaces

Mohawk Industries, Inc.
P.O. Box 71
Adams, MA 01220
 Tempwood wood and coal stoves

Multi-Fuel Energy Systems
2185 N. Sherman Dr.
Indianapolis, IN 46218
 Dual fuel furnaces

National Stove Works, Inc.
Box 640
Cobleskill, NY 12043
 Thermo-Control stoves; furnaces

New Hampshire Wood Stoves, Inc.
38 Haywood St.
Greenfield, MA 01301
 Air-tight wood stoves

Preston Distributors
Whidden St.
Lowell, MA 01852
   Wood and coal stoves

Riteway Manufacturing
Box 6
Harrisonburg, VA 22801
   Dual fuel furnaces; wood stoves

Superior Fireplace Co.
4325 Artesia Ave.
Fullerton, CA 92633
   Heatform circulating fireplaces

Therm-Kon Products
207 E. Mill Road
Galesville, WI 54630
   Wood-coal stoves and furnaces

U.S. Stove Co.
Box 151
South Pittsburg, TN 37380
   Wood stoves

Vega Industries, Inc.
Mt. Pleasant, IA 52644
   Heat circulating fireplaces

CHIMNEY CLEANING PRODUCTS

Anachron
Box 8860
Portland, OR 97208
   Chemical flue cleaner

Anchor Tools and Wood Stoves
618 N.W. Davis
Portland, OR 97209
   Chimney brushes

Preston Distributors
14 Whidden St.
Lowell, MA 01852
   Chemical flue cleaner

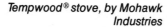

Tempwood® stove, by Mohawk Industries

SOLAR COLLECTORS

Applied Energy Systems
3007 E. Jackson St.
Broken Arrow, OK 74012
  Concentrating collectors

Colorado Sunworks
P.O. Box 455
Boulder, CO 80306
  Collectors

Fal Bel Energy Systems
Box 6
Greenwich, CT 06830
  Flat plate collectors

Home Spun Sun Works
Box 229
Cazadero, CA 95421
  Flat plate collectors

International Solarthermics Corp.
Box 397
Nederland, CO 80466
  Self-contained detached system

Solar State Systems
2821 Ladybird Lane
Dallas, TX 75220
  Collectors; controllers

Sun-Wall, Inc.
Box 9723
Pittsburg, PA 15229
  Flat plate collectors

Sunworks, Inc.
669 Boston Post Road
Guilford, CT 06437
  Flat plate collectors

Thomason Solar Homes, Inc.
6802 Walker Mill Rd., S.E.
Washington, DC 20027
  Solaris fluid systems

Ying Manufacturing Corp.
1957 W. 144th St.
Gardena, CA 90249
  Flat plate collectors

Grundfos booster pump

SOLAR HOT WATER

A. O. Smith Corp.
P.O. Box 28
Kankakee, IL 60901
Solar water heaters

Beutel Solar Heater Co.
1527 N. Miami Ave.
Miami, FL 33132
Solar water heater

Grundfos Pumps Corp.
2555 Clovis Ave.
Clovis, CA 93612
Hot water pumps

Rheem Water Heater Div.
City Investing Company
7600 S. Kedzie Ave.
Chicago, IL 60652
Solar water heaters

Solar Components
Div. of Kalwall Corp.
P.O. Box 237
Manchester, NH 03105
Tanks, collectors

Solar Energy Digest
P.O. Box 17776
San Diego, CA 92117
Solapod water heaters

Solar Energy Systems
5825 Green Ridge Dr., N.E.
Atlanta, GA 30328
Pumps and controllers

LIGHT TRANSMITTING PANELS

American Acrylic Corp.
173 Marine St.
Farmingdale, NY 11735

Barclay Manufacturing Co.
65 Industrial Road
Lodi, NJ 07644

Berdon, Inc.
711 Olympic Blvd.
Santa Monica, CA 90401

Fiberglass Plastics, Inc.
7395 N. W. 34th Ct.
Miami, FL 33147

Glasteel, Inc.
1727 Buena Vista St.
Duarte, CA 91010

Idaho Chemical Industries
P.O. Box 7866
Boise, ID 83707

Kallwall Corp.
1111 Candia Road
Manchester, NH 03103

Lasco Industries
1561 Chapin Road
Montebello, CA 90640

Reichold Chemicals
P.O. Box 7
Grand Junction, TN 38039

Resolite Division
Route 19
Zelienople, PA 16063

Thorolyte Fiberglass, Inc.
8969 S.E. 58th St.
Portland, OR 97206

Transparent Products Corp.
1727 W. Pico Blvd.
Los Angeles, CA 90015

Vistron Corp, Filon Div.
12333 S. Van Ness St.
Hawthorne, CA 90250

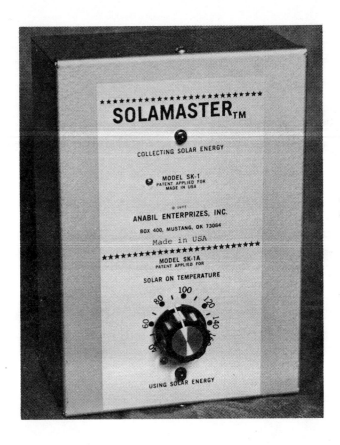

Solamaster controller, by Anabil Enterprizes

GREENHOUSE EQUIPMENT

Edward Owen Engineering
Snow Shoe, PA 16874
   Greenhouse kits

Lord & Burnham
Irvington, NY 10533
   Attached greenhouses

Pacific Coast Greenhouse Mfg. Co.
8360 Industrial Ave.
Cotati, CA 94928
   Greenhouses; equipment

Solar Sauna
Box 466
Hollis, NH 03049
   Greenhouse, sauna kits

Vegetable Factory
Box 2235 GCS
New York, NY 10017
   Solar greenhouses

BLOWERS AND FANS

Contemporary Systems, Inc.
68 Charlonne St.
Jaffrey, NH 03452
   Solar collector fans

Kool-O-Matic
1831 Terminal Road
Niles, MI 49120
   Powered attic ventilators

Sun Stone Solar Energy Equip.
Box 941
Sheboygan, WI 53081
   Blowers, pumps

Surplus Center
1000 West "O" St.
Lincoln, NE 68501
   Parts for fans; water and wind equipment

Interior of a Davis Caves underground home

CONTROLLERS AND INSTRUMENTS

Ammark Corp.
12-22 River Road
Fairlawn, NJ 07410
  Controllers; set-back thermostats

Anabil Enterprizes, Inc.
525 S. Aqua Clear Dr.
Mustang, OK 73064
  Differential solar controllers

B. E. S. T., Inc.
Rt. 1, Box 106
Necedah, WI 54646
  1kw to 5kw solid state inverters

Dalen Products, Inc.
201 Sherlake Dr.
Knoxville, TN 37922
  Thermal motors

F. W. Dwyer
Michigan City, IN 46360
  Hand held anemometers

Heat Motors, Inc.
635 W. Grandview Ave.
Sierra Madre, CA 91024
  Thermal motors

Heliotrope General
3733 Kenora Dr.
Spring Valley, CA 92077
  Controllers; motorized dampers

L. W. Gay Stoveworks
156 Vernon Rd.
Brattleboro, VT 05301
  Bimetal damper thermostats

Natural Power, Inc.
New Boston, NH 03070
  Differential thermostats

R. A. Simerl Instruments
238 W. St.
Annapolis, MD 21401
  Anemometers

Real Gas & Electric Co.
278 Borham Ave.
Santa Rosa, CA 95402
  Synchronous inverters

Solar Control Corp.
5721 Arapahoe Rd.
Boulder, CO 80303
  Solar controllers

Spectrex Co.
Bragg Hill Rd.
Waitsfield, VT 05673
  Electrical meters

Trade Winds Instruments
10823 12th Ave., N.E.
Seattle, WA 98125
  Anemometers

OTHER SOLAR EQUIPMENT

Arkla Industries, Inc.
950 E. Virginia St.
Evansville, IN 47701
  Absorption solar cooler

Artech Corp.
2816 Marine St.
Farmingdale, NY 11735
  Change-of-state storage salts

Edmund Scientific
1875 Edscorp Bldg.
Barrington, NJ 08007
  Photovoltaic cells

Solar, Inc.
Box 246
Mead, NE 68024
  Eutectic salts solar heat storage

EARTH-SHELTERED CENTER

Davis Caves, Inc.
P.O. Box 102
Armington, IL 61721
  "Earth-Powered" underground homes;
  engineering and construction.

Gar-San Enterprises, Inc.
544 3rd St.
Elk River, MN 55330
  Earth-contact and underground homes.

HYDROELECTRIC

Independent Power Developers
Box 1467
Noxon, MT 59853
  High and low head hydro-turbines.

Short Stopper Electric
Route 4, Box 471B
Coos Bay, OR 97420
  High and low head hydro-turbines

Small Hydroelectric Systems and Equipment
15220 S.R. 530
Arlington, WA 98223
  *Peltech impulse turbines, 6" to 39", for high-head
  installations. Bronze, aluminum, steel castings. Site
  engineering assistance.*

The James Leffel and Company
Springfield, OH 45501
  *Hydro-turbines of all sizes; custom castings. Hoppes
  self-contained turbine and generator units, in sizes
  from 1 kw to 10 kw. Engineering assistance.*

WIND-POWERED GENERATORS AND EQUIPMENT

Aeolian Energy
Rt. 4
Ligonier, PA 15658
  Four blade turbine generates 8 kw at 18 m.p.h.

American Wind Turbine, Inc.
1016 E. Airport Rd.
Stillwater, OK 74074
  *High-speed turbine (48-blade) develops 2 kw in 20
  m.p.h. wind.*

Automatic Power, Inc.
Pennwalt Corp.
205 Hutcheson St.
Houston, TX 77003
  *French-built Aerowatt wind generators from 24 to
  4,100 watts.*

Dynergy Corp.
Box 428
Laconia, NH 03426
  *15 ft. (4.6 meter) diameter Darrieus vertical-axis tur-
  bine develops 3,300 watts at 24 m.p.h.*

Peltech impulse turbines

Entertech Corp.
P.O. Box 420
Norwich, VT 05055
   Entertech 1500 is a 13.2 ft. (4 meter) diameter machine
that drives an induction generator to develop 1500
watts at 22 m.p.h.

Grumman Energy Systems
4175 Veterans Memorial Hwy.
Ronkondoma, NY 11779
   3 bladed, 25 ft. (7.62 meter) diameter rotor develops
20 kw at 29 m.p.h.

Millville Windmills
P.O. Box 32
Millville, CA 96062
   3 bladed, 25 ft. diameter machine develops 10 kw at
25 m.p.h.

Pinson Energy Corp.
P.O. Box 7
Marston's Mills, MA 02648
   Vertical axis "Cycloturbine" 12 ft. diameter rotor
develops 2000 watts at 24 m.p.h.

Sencenbaugh Wind Electric
P.O. Box 11174
Palo Alto, CA 94306
  12 foot diameter, 3 bladed machine develops
  1000 watts at 23 m.p.h.

TRW Enterprises
72 W. Meadow Lane
Sandy, VT 04070
  Vertical axis wind turbines and generators,
  1 kw to 5 kw.

Unarco-Rohn
6718 W. Plank Rd.
Peoria, IL 61656
  Towers for wind machines.

Wind Engineering Corp.
Box 5936
Lubbock, TX 79417
  Windgen 25 generator develops 25 kw

Wind Power Systems, Inc.
P.O. Box 17323
San Diego, CA 92117
  3 bladed, 32.8 ft. (10 meter) turbine develops
  6 kw at 18 m.p.h.

Dynergy's 5-meter Darrieus rotor

INDEX

Other Successful Books...

BOOK OF SUCCESSFUL FIREPLACES, 20th ed., Lytle. The expanded, updated edition of the book that has been a standard of the trade for over 50 years—over a million copies sold! Advice is given on selecting from the many types of fireplaces available, on planning and adding fireplaces, on building fires, on constructing and using barbecues. Also includes new material on wood as a fuel, woodburning stoves, and energy savings. 8½″ x 11″; 128 pp; over 250 photos and illustrations. $5.95 Paper.

SUCCESSFUL ROOFING & SIDING, Reschke. "This well-illustrated and well-organized book offers many practical ideas for improving a home's exterior." *Library Journal.* Here is full information about dealing with contractors, plus instructions specific enough for the do-it-yourselfer. All topics, from carrying out a structural checkup to supplemental exterior work like dormers, insulation, and gutters, fully covered. Materials to suit all budgets and home styles are reviewed and evaluated. 8½″ x 11″; 160 pp; over 300 photos and illustrations. $5.95 Paper. (Main selection Popular Science and McGraw-Hill Book Clubs)

SUCCESSFUL HOME ADDITIONS, Schram. For homeowners who want more room but would like to avoid the inconvenience and distress of moving, three types of home additions are discussed: garage conversion with carport added; bedroom, bathroom, sauna addition; major home renovation which includes the addition of a second-story master suite and family room. All these remodeling projects have been successfully completed and, from them, step-by-step coverage has been reported of almost all potential operations in adding on to a home. The straightforward presentation of information on materials, methods, and costs, as well as a glossary of terms, enables the homeowner to plan, arrange contracting, or take on some of the work personally in order to cut expenses. 8½″ x 11″; 144 pp; over 300 photos and illustrations. Cloth $12.00. Paper $5.95.

FINISHING OFF, Galvin. A book for both the new-home owner buying a "bonus space" house, and those who want to make use of previously unused areas of their homes. The author advises which jobs can be handled by the homeowner, and which should be contracted out. Projects include: putting in partitions and doors to create rooms; finishing off floors and walls and ceilings; converting attics and basements; designing kitchens and bathrooms, and installing fixtures and cabinets. Information is given for materials that best suit each job, with specifics on tools, costs, and building procedures. 8½″ x 11″; 144 pp; over 250 photos and illustrations. Cloth $12.00. Paper $5.95.

SUCCESSFUL FAMILY AND RECREATION ROOMS, Cornell. How to best use already finished rooms or convert spaces such as garage, basement, or attic into family/recreation rooms. Along with basics like lighting, ventilation, plumbing, and traffic patterns, the author discusses "mood setters" (color schemes, fireplaces, bars, etc.) and finishing details (flooring, wall covering, ceilings, built-ins; etc.) A special chapter gives quick ideas for problem areas. 8½″ x 11″; 144 pp; over 250 photos and diagrams. (Featured alternate for McGraw-Hill Book Clubs) $12.00 Cloth. $5.95 Paper.

SUCCESSFUL LOG HOMES, Ritchie. Log homes are becoming increasingly popular—low cost, ease of construction and individuality being their main attractions. This manual tells how to work from scratch whether cutting or buying logs—or how to remodel an existing log structure—or how to build from a prepackaged kit. The author advises on best buys, site selection, evaluation of existing homes, and gives thorough instructions for building and repair. 8½″ x 11″; 168 pp; more than 200 illustrations including color. $12.00 Cloth. $5.95 Paper.

SUCCESSFUL SMALL FARMS — BUILDING PLANS & METHODS, Leavy. A comprehensive guide that enables the owner of a small farm to plan, construct, add to, or repair buildings at least expense and without disturbing his production. Emphasis is on projects the farmer can handle without a contractor, although advice is given on when and how to hire work out. Includes basics of farmstead layout, livestock housing, environmental controls, storage needs, fencing, building construction and preservation, and special needs. 8½″ x 11″; 192 pp; over 250 illustrations. $14.00 Cloth. $5.95 Paper.

SUCCESSFUL HOME REPAIR — WHEN NOT TO CALL THE CONTRACTOR. Anyone can cope with household repairs or emergencies using this detailed, clearly written book. The author offers tricks of the trade, recommendations on dealing with repair crises, and step-by-step repair instructions, as well as how to set up a preventive maintenance program. 8½″ x 11″; 144 pp; over 150 illustrations. $12.00 Cloth. $5.95 Paper.

Available from your local bookseller or order direct from:
STRUCTURES PUBLISHING COMPANY • BOX 1002 • FARMINGTON, MICHIGAN 48024
(Please add $1.00 postage and handling for mail orders)

JA